T0345007

Twisted Logic

Twisted Logic: Puzzles, Paradoxes, and Big Questions delves into the intriguing world of twisted logic, where everyday conundrums, bewildering paradoxes, and life's big questions are investigated and decoded. Crafted for the curious mind, this book sheds light on how our intuition and common sense can often mislead us. Without the need for technical jargon or mathematical prowess, it serves as your personal compass through fascinating intellectual landscapes and ultimate explorations. From the quirky corners of Bayesian reasoning to practical strategies in daily choices, this is your companion for a clearer way of thinking.

Features:

- A comprehensive toolkit to refine your cognitive processes and avoid common pitfalls. Insights into the oddities of probability, strategy, and fate that govern our lives.
- A fresh perspective on everyday decisions and life's larger dilemmas, including finding everything from a place to eat to a new home to a life partner.
- Practical advice on optimising daily routines, such as determining the best time of day to arrange important appointments.
- Thought-provoking 'When Should We?' questions that challenge us to think critically about decision-making in our lives.
- Prepare to challenge your perceptions and unveil hidden truths. *Twisted Logic* is an enlightening adventure that promises to transform the mundane into the extraordinary.
- **Embark on a journey where the only thing certain is the thrill of the unknown.**

Leighton Vaughan Williams, PhD, is not only a distinguished professor but also a storyteller who bridges the worlds of academia and everyday life. His work at Nottingham Business School as a teacher and leader of its research and forecasting units uniquely equips him to navigate you through intriguing puzzles, paradoxes, and profound questions with both wit and accessible wisdom.

Twisted Logic
Puzzles, Paradoxes, and Big Questions

Leighton Vaughan Williams

CRC Press
Taylor & Francis Group
Boca Raton London New York

CRC Press is an imprint of the
Taylor & Francis Group, an **informa** business

A CHAPMAN & HALL BOOK

First edition published 2025
by CRC Press
2385 NW Executive Center Drive, Suite 320, Boca Raton FL 33431

and by CRC Press
4 Park Square, Milton Park, Abingdon, Oxon, OX14 4RN

CRC Press is an imprint of Taylor & Francis Group, LLC

© 2025 Leighton Vaughan Williams

Reasonable efforts have been made to publish reliable data and information, but the author and publisher cannot assume responsibility for the validity of all materials or the consequences of their use. The authors and publishers have attempted to trace the copyright holders of all material reproduced in this publication and apologize to copyright holders if permission to publish in this form has not been obtained. If any copyright material has not been acknowledged please write and let us know so we may rectify in any future reprint.

Except as permitted under U.S. Copyright Law, no part of this book may be reprinted, reproduced, transmitted, or utilized in any form by any electronic, mechanical, or other means, now known or hereafter invented, including photocopying, microfilming, and recording, or in any information storage or retrieval system, without written permission from the publishers.

For permission to photocopy or use material electronically from this work, access www.copyright. com or contact the Copyright Clearance Center, Inc. (CCC), 222 Rosewood Drive, Danvers, MA 01923, 978-750-8400. For works that are not available on CCC please contact mpkbookspermissions@ tandf.co.uk

Trademark notice: Product or corporate names may be trademarks or registered trademarks and are used only for identification and explanation without intent to infringe.

ISBN: 9781032515731 (hbk)
ISBN: 9781032513348 (pbk)
ISBN: 9781003402862 (ebk)

DOI: 10.1201/9781003402862

Typeset in Times
by codeMantra

Dedicated to:

Mum, Dad, and my wife, Julie

Contents

Preface

THE WONDERFUL WORLD OF CHANCE, LOGIC, AND INTUITION

Welcome to the delicate dance between logic and intuition, to the dynamics of chance and strategic thinking. *Twisted Logic* invites you on a journey to apply these concepts across various aspects of life, making them both accessible and captivating.

A UNIVERSAL INVITATION

Regardless of your background or interests, this book has something for you. The content is tailored to ensure that everyone can find value and insight.

EMPOWERING YOUR LOGICAL TOOLKIT

Dive into the art of clear thinking with an exploration of common cognitive errors. This book is your arsenal for honing logical reasoning, equipped with practical tools to navigate life's challenges more effectively.

THE THRILL OF DISCOVERY THROUGH BAYESIAN REASONING

Prepare to be amazed by the counter-intuitive insights of Bayesian reasoning. This section will guide you through its unexpected conclusions, showcasing its power in transforming how you interpret the world.

PUZZLES, PARADOXES, AND PRACTICAL WISDOM

Embark on a riveting adventure filled with classic and novel puzzles and paradoxes. Each chapter is designed to challenge your assumptions and introduce you to new ways of problem-solving, touching upon the oddities and enigmas of making decisions.

REALITY VERSUS INTUITION: A SURPRISING EXPLORATION

Our intuitive judgments are often put to the test by reality's unpredictability. This section delves into real-world scenarios, revealing the gaps in our intuitive logic and offering strategies to bridge them.

UNVEILING LIFE'S MYSTERIES AND STRATEGIES

Discover the unexpected connections that influence daily decisions and life's larger strategies, from finding the best dining spots to navigating the complexities of relationships. This exploration is about unveiling the wisdom hidden in plain sight.

EXPLORING GAMES OF CHANCE AND SKILL

Gaining an edge at games involving chance and skill might seem elusive, but look no further. Break down the myths and arm yourself with strategies for games from Blackjack to Rock-Paper-Scissors.

A FRESH LOOK AT NUMBERS AND NATURE

Identify the numerical patterns that underpin the natural world and how they can reveal truths or deceit. This section encourages a deeper appreciation for the numbers that shape our understanding of the world.

CHALLENGING CONVENTIONS AND UNCOVERING TRUTHS

Question everything. From gut feelings to widely accepted beliefs, learn why our perceptions often mislead us and how to see beyond them.

EXTRAORDINARY INSIGHTS INTO ORDINARY THINGS

Investigate the quirks of finance, sports, and the unpredictability of life itself. From the impact of the seasons on your investments to the science behind winning, you're covered.

CONTEMPLATING EXISTENCE AND EVERYDAY DECISIONS

From philosophical ponderings about humanity's future to practical advice on scheduling appointments, this book doesn't shy away from the big or small questions. Get ready for deep thoughts and practical tips alike.

EMBARK ON THIS JOURNEY

Twisted Logic makes the complex comprehensible and the ordinary fascinating. With each page, you'll find your perceptions challenged, your knowledge expanded, and your curiosity satisfied. Join this enlightening expedition as we demystify the world through logic, humour, and unexpected revelations.

The Author

Professor Leighton Vaughan Williams, PhD, is a renowned teacher, researcher, and author with a passion for making complex concepts accessible to all. As an esteemed faculty member at Nottingham Business School, Leighton has successfully merged academic theories with real-life applications, making significant contributions to the understanding of probability, risk, and decision-making under uncertainty.

Throughout his career, Leighton has provided expert evidence in prestigious legal settings, both nationally and internationally, and before legislative and regulatory bodies, including committees of both the House of Commons and House of Lords. His advisory expertise has informed policy decisions for various Government departments, translating sophisticated research into actionable strategies.

Globally recognised also for his teaching, Leighton has turned mathematical and statistical principles into accessible wisdom, resolving everyday dilemmas and unravelling paradoxes for students around the world. His leadership at international conferences and prolific contributions to scholarly literature have established him as a pioneer in bridging theory with real-world applications.

His commitment to practical understanding is epitomised in his acclaimed book, *Probability, Choice, and Reason*, published by Chapman & Hall/CRC in 2021, which synthesises scholarly insights with hands-on guidance.

Leighton was educated at University College, Cardiff, graduating summa cum laude, and at Hertford College, Oxford University.

A Question of Evidence

1

When Should We Update Our Beliefs?

Imagine emerging from a cave for the first time and watching the sun rise. You have never witnessed this before, and in this thought experiment, you are unable to tell whether it's a regular occurrence, an infrequent event, or a once-in-a-lifetime happening.

As each day passes, however, you observe the dawn again and again: you gradually grow to expect it. With each sunrise, you become more confident that this is a regular event. With this growing confidence, you forecast that the sun will rise again the next day.

This is an illustration of what so-called Bayesian reasoning is about. Bayes' theorem is a tool that allows us to adjust our understanding of the world based on our observations over time. It represents a process of continuous learning and understanding, pushing us gradually nearer to the truth as we are exposed to more experiences and to more information.

That's the essence of Bayesian reasoning: adjusting our beliefs based on new information.

THE BIRTH OF BAYESIAN THINKING

The Bayesian perspective on the world can be traced to the Reverend Thomas Bayes, an 18th-century clergyman, statistician, and philosopher. The Bayesian approach advocated predicting future events based on past experiences. His ideas were in a fundamental sense different from the prevailing philosophical ideas of his time, notably those of Enlightenment philosopher David Hume.

Hume argued that we should not justify our expectations about the future based on our experiences of the past, because there is no law stating that the future will always mirror the past. As such, we can never be certain about our knowledge derived from experience. For Hume, therefore, the fact that the sun had risen every day up to now was no guarantee that it would rise again tomorrow. In contrast, Bayes provided a tool for

DOI: 10.1201/9781003402862-1

predicting the likelihood of such events based on past experiences and observations. His method can be applied consistently to the sciences, social sciences, and many aspects of our everyday lives.

Unlike the philosopher David Hume, who argued that past experiences don't guarantee future outcomes, Bayes focused on how we can use past events to predict the likelihood of future ones. Bayes' approach is not just academic; it's a practical tool.

BAYES' THEOREM: AN EVERYDAY TOOL FOR REFINING PREDICTIONS

So how does what is known as Bayes' theorem help us in our everyday lives and beyond? As it turns out, it's an important way of helping us to refine our belief of what is true and what is false. Let's look more closely into this by breaking Bayes' theorem down into its key components:

1. **Establish a Prior Hypothesis:** The starting point in Bayesian reasoning involves the establishment of an initial hypothesis, which may or may not be true. This hypothesis, also known as the 'prior' belief or 'prior probability' that you assign to this belief being true, is based on the information available to you. For instance, if you're trying to predict whether it will rain tomorrow, you might estimate the initial likelihood (or 'prior probability') based on your personal observation of current weather patterns or conditions.
2. **Observe New Evidence:** Once you establish a prior probability, you'll then need to consider updating this when any new information becomes available. In the weather example, evidence could be anything from new dark clouds gathering or else dispersing to a sudden rise or drop in temperature.
3. **Assess to What Extent This New Evidence Is Consistent with Your Initial Hypothesis:** Bayesian reasoning doesn't stop at just gathering evidence. It also involves considering evidence that is consistent with, or inconsistent with, your initial hypothesis. For example, if there is an increase or decrease in wind speed, this might be considered additional evidence that you should take into account in estimating the probability of rain.

Let's break down again how Bayes' theorem helps us refine our beliefs:

1. **Establishing a Starting Point (The Prior Hypothesis):** Imagine you're trying to predict if it will rain tomorrow. Your 'prior hypothesis' is your initial estimate, based on what you currently know about the weather conditions.
2. **Incorporating New Information (New Evidence):** Now, suppose you observe unexpected dark clouds gathering in the sky. This new information should logically influence your prediction about the weather.
3. **Combining Old and New Insights (Assessing Consistency):** Bayesian reasoning involves integrating the new evidence with your initial estimate. You assess whether the appearance of dark clouds increases the likelihood of rain tomorrow.

By applying Bayes' theorem, you adjust your belief based on the new evidence. If dark clouds often lead to rain, you increase your belief that it will rain. If not, you adjust accordingly.

Visualising Bayes' Theorem

Think of Bayes' theorem as a formula that combines your initial estimate with new information to give you a better estimate.

Beyond Weather: The Broad Applications of Bayes' Theorem

Bayesian reasoning isn't just about predicting the weather. It's used in medicine to interpret test results, in finance to assess investment risks, in sports for game strategies, and so on. It's a tool that refines our understanding, helping us make more informed decisions.

HOW BAYES' THEOREM ALLOWS US TO UPDATE OUR BELIEFS

In essence, Bayes' theorem permits us to establish an initial hypothesis, and to enter any supportive and contradicting evidence into a formula which can be used to update our belief in the likelihood that the hypothesis is true.

Consider a scenario where we evaluate our initial hypothesis. For simplicity, we label the probability that this hypothesis is correct as 'a'. This probability is our starting point, reflecting our initial estimate based on prior knowledge or assumptions before encountering new data.

Next, we introduce 'b', which represents the likelihood that some new evidence we come across is consistent with our initial hypothesis being true. This is a critical element of Bayesian updating.

Conversely, 'c' is used to denote the probability of observing the same new evidence but under the condition that our initial hypothesis is false. This estimate is equally essential because it helps us understand the significance of the evidence in the context of our hypothesis not being true.

With these definitions in place, Bayes' Theorem provides a powerful formula: Revised (posterior) probability that our initial hypothesis is correct= $ab/[ab+c(1-a)]$

This formula is a mathematical tool that updates our initial belief 'a' in light of the new evidence.

The result is an updated (or 'posterior') probability that reflects a more informed stance on the initial hypothesis.

This process, termed Bayesian updating, is a methodical approach that enables us to refine our beliefs incrementally. As we gather more evidence, we iteratively apply this updating process, allowing our beliefs to evolve closer to reality with each new piece of information. This ongoing refinement is a cornerstone of the Bayesian approach, emphasising the importance of evidence in shaping our understanding and beliefs.

BAYES' THEOREM: A POWERFUL TOOL

Bayes' theorem offers us a weapon against biases in our intuition, which can often mislead us. For example, intuition can sometimes lead us to ignore previous evidence or to place too much weight on the most recent piece of information. Bayes' theorem offers a roadmap that assists us in balancing the weight of previous and new evidence correctly. In this way, it provides a method for us to fine-tune our beliefs, leading us gradually closer to the truth as we gather and consider each new piece of evidence.

CONCLUSION: THE BAYESIAN BEACON

Bayes' theorem is more than a mathematical concept; it's a guide through the uncertain journey of life. It teaches us to be open to new information and to continually adjust our beliefs. From daily decisions like weather predictions to complex scientific theories, Bayes' theorem is a bridge from uncertainty to better understanding, helping us navigate life's puzzles with more confidence and precision.

It does so in a structured way, dealing with new evidence, guiding us gradually to more informed beliefs. It encourages us always to be open to new evidence and to adjust our beliefs and expectations accordingly. Bayes' theorem is in this sense a master key to understanding the world around us.

When Should We Believe a Friend?
Exploring the Reliability of Evidence

UNDERSTANDING EVIDENCE THROUGH A STORY

Imagine a situation where your friend, known for her upstanding character, is accused of vandalising a shop window. The only evidence against her is a police officer's identification. You know her well and find it hard to believe she would commit such an act.

SETTING THE STAGE FOR BAYESIAN ANALYSIS

In Bayesian terms, the 'condition' is your friend being accused, while the 'hypothesis' is that she's guilty. To apply Bayes' theorem, we consider three probabilities:

1. **Prior Probability:** Based on your knowledge of your friend, you might initially think there's a low chance of her guilt, say 5%. This is your 'prior'—your belief before considering the new evidence.
2. **Likelihood of Evidence If Guilty:** Consider how reliable the officer's identification is. If your friend were guilty, what's the chance that the officer would identify her correctly? Let's estimate this at 80%.
3. **Likelihood of Evidence If Innocent:** What's the chance of the officer mistakenly identifying your friend if she's innocent? Factors like similar appearances or biases could play a role. Let's estimate this at 15%.

THE ITERATIVE NATURE OF BAYESIAN UPDATING

Bayes' theorem allows for continual updates. If new evidence arises, you can recalculate, using your updated belief as the new 'prior'. This process offers a dynamic way to assess situations as they evolve.

WHEN EVIDENCE DOESN'T ADD UP

In cases where evidence is equally likely whether the hypothesis is true or false, it doesn't change our belief. It's crucial to evaluate the quality of evidence, not just its existence.

CHALLENGES IN ASSIGNING PROBABILITIES

While assigning precise probabilities to real-life situations can be challenging, the exercise is invaluable. It forces us to think critically and systematically about our beliefs and how new information affects them.

The Unfolding Story

Now let's consider the story in a little more detail. You've received a phone call from your local police station. An officer tells you that your friend, someone you've known for years, is currently assisting the police in their investigation into a case of vandalism. The crime in question involves a shop window that was smashed on a quiet street, close to where she resides. Furthermore, the incident took place at noon that day, which happens to be her day off work.

You had heard about the incident, but had no reason to believe your friend was involved. After all, she's not a person known for reckless or unlawful behaviour.

However, this is where the narrative takes a twist. Your friend comes to the phone and tells you that she's been charged with the crime. The accusation primarily stems from the assertion of a police officer who has positively identified her as the offender. There's no other evidence, such as CCTV footage or eyewitness testimonies, to substantiate the officer's claim.

She vehemently maintains her innocence, insisting it's a case of mistaken identity.

The Challenge

Now, as a follower of Bayes as well as being a close friend, you find yourself in a position where you need to evaluate the probability that she has committed the crime before deciding how to advise her. This challenge leads us to the central theme of this section—the application of Bayes' theorem to real-life situations.

Before we proceed, let's clarify our terms. The 'condition' in this context is that your friend has been accused of causing the criminal damage. The 'hypothesis' we aim to assess is the probability that she is indeed guilty.

Bayes' Theorem and Its Application

So, how does Bayes' theorem help us answer this question? Well, Bayes' theorem is a formula that describes how to update the probabilities of hypotheses being true when given new evidence. It follows the logic of probability theory, adjusting initial beliefs based on the weight of evidence.

To apply Bayes' theorem, we need to estimate three crucial probabilities:

1. **Prior probability ('a')**

   The prior probability refers to the initial assessment of the hypothesis being true, independent of the new evidence. In this scenario, it equates to the likelihood you assign to your friend being guilty before you hear the evidence.

   Considering you've known her for years and her involvement in such an act is uncharacteristic, you might deem this probability low. After a thoughtful consideration of your friend's past actions and character, allowing for the fact that she was off work on that day and in the neighbourhood, let's say you assign a 5% chance (0.05) to her being guilty.

   Assigning this prior probability requires an honest evaluation of your initial beliefs, unaffected by the newly received information.

2. **Conditional probability of evidence given hypothesis is true ('b')**

   Next, you need to estimate the likelihood that the new evidence (officer's identification) would have arisen if your friend were indeed guilty.

   This estimate might be guided by factors such as the officer's reliability, credibility, and proximity to the crime scene. For the sake of argument, let's estimate this probability to be 80% (0.8).

3. **Conditional probability of evidence given hypothesis is false ('c')**

   The third estimate involves figuring out the probability that the new evidence would surface if your friend is innocent. This entails gauging the chance that the officer identifies your friend as the offender when she isn't guilty.

   The probability could be influenced by several factors—perhaps the officer saw someone of similar age and appearance, jumped to conclusions, or has other motivations. For the purposes of our discussion, let's estimate this probability to be 15% (0.15).

Probabilities Adding Up

An interesting point to note is that the sum of probabilities 'b' and 'c' doesn't necessarily have to equal 1. Just for example, the police officer might have a reason to identify your friend either way (whether she's guilty or innocent), in which case the sum of 'b' and 'c' could exceed 1. Alternatively, the officer may be reluctant to positively identify a suspect in any circumstance unless he is absolutely certain; in which case b plus c may well sum to rather less than 1. In this particular narrative, b plus c add up to 0.95.

Calculation and Interpretation

With these estimates in hand, we can now apply Bayes' theorem, which calculates the posterior probability (the updated probability of the hypothesis being true after considering new evidence) using the formula: $ab/[ab+c\,(1-a)]$.

In our case, substituting the values results in a posterior probability of around 21.9%. What does this mean? Despite the officer's confident identification (a seemingly strong piece of evidence), there's only a 21.9% probability that your friend is guilty given the current information.

This result may seem counterintuitive. However, this discrepancy arises from our understanding of prior probability and the weight we assign to the new evidence. We must remember that the officer's identification is only one piece of the puzzle, and its strength as evidence is balanced against the prior probability and the potential for a false identification.

Updating the Probability

The beauty of Bayes' theorem lies in its iterative nature. Let's suppose that another piece of evidence emerges—say, a second witness identifies your friend as the culprit. You can reapply Bayes' theorem, using the posterior probability from the previous calculation as the new prior probability. This iterative process allows you to incorporate additional pieces of evidence, each of which updates the probability you assign to your friend's guilt or innocence.

Cases Where Evidence Adds No Value

Consider a situation where 'b' equals 1 and 'c' also equals 1. This would imply that the officer would identify your friend as guilty whether she was or not. In such cases, the identification fails to update the prior probability, and the posterior probability remains the same as the initial prior probability.

The Imperfections of Assigning Probabilities

Now, it's worth recognising the potential difficulty in assigning precise probabilities to real-life situations. After all, our scenario involves complex human behaviour and a unique event.

However, our inability to determine precise probabilities shouldn't lead us to dismiss the process. In fact, this process of estimation is what we're doing implicitly when we evaluate situations in our everyday lives.

While the results might not be perfect, Bayes' theorem provides a systematic approach to updating our beliefs in the face of new evidence.

CONCLUSION: BAYESIAN REASONING IN REAL LIFE

Bayes' theorem provides a structured approach to incorporating new evidence into our beliefs. It's a tool that enhances our decision-making, offering a mathematical framework to navigate uncertainties, from everyday dilemmas to complex legal and medical decisions.

As we grapple with uncertainty, the application of Bayes' theorem allows us to transition from ignorance to knowledge, systematically and rationally. Thus, whether we're faced with a shattered shop window or any other challenging situation, we have a powerful tool to help us navigate our path towards truth.

When Should We Believe the Eyewitness? Exploring the Taxi Problem

THE BASICS OF THE TAXI PROBLEM

Let's set the stage for our story. We're in New Brighton, a city with a fleet of 1,000 taxis. Of these, 850 are blue and 150 are green. One day, a taxi is involved in an accident with a pedestrian and leaves the scene. We don't know the colour of the taxi, and we don't have any reason to believe that blue or green taxis are more likely to be involved in such incidents.

An independent eyewitness now comes forward. She saw the accident and claims that the taxi was green. To verify the reliability of her account, investigators conduct a series of observation tests designed to recreate the conditions of the incident. These tests reveal that she is correct about the colour of a taxi in similar conditions 80% of the time.

So, what is the likelihood that the taxi involved was actually green?

INITIAL PROBABILITIES AND INTUITIVE ESTIMATES

Your first instinct might be to believe that the chance that the taxi was green is around 80%. This assumption is based on the witness's track record of identifying the colour of a taxi accurately. However, this conclusion doesn't consider other crucial information—the overall number of blue and green taxis in the city.

Given the total taxi population, only 15% of them are green (150 out of 1,000), while a substantial 85% are blue. Ignoring this 'base rate' of taxi colours leads to a common mistake known as the 'Base Rate Fallacy'.

APPLYING BAYES' THEOREM TO THE TAXI PROBLEM

Bayes' theorem is a method that helps us adjust our initial estimates based on new evidence but allowing for this base rate of the total numbers of blue and green taxis. In this way, it offers a means of updating our initial estimates after taking account of some new evidence.

For our Taxi Problem, the new evidence is the witness statement. The witness says the taxi was green, and we know that there's an 80% chance that she is correct if the taxi was indeed green (based on her observation test). But there's also a 20% chance that she would mistakenly say the taxi was green if it were blue.

Bayes' theorem helps us adjust initial beliefs with new evidence, considering the base rate. Here's how it works in the Taxi Problem:

1. **Prior Probability:** Initially, there's only a 15% chance (150 out of 1,000 taxis) that the taxi is green.
2. **Conditional Probability of Green Taxi (If Witness Correct):** The eyewitness is correct 80% of the time.
3. **Conditional Probability of Green Taxi (If Witness Incorrect):** There's a 20% chance the eyewitness would mistakenly identify a blue taxi as green.

After applying Bayes' theorem, the adjusted (or 'posterior') probability that the taxi is green is just 41%, using the formula: $ab/[ab+c(1-a)]$.

THE ROLE OF NEW EVIDENCE AND MULTIPLE WITNESSES

What happens if another eyewitness comes forward? Suppose this second witness also reports that the taxi was green and, after a similar set of tests, is found to be correct 90% of the time. Now we should recalculate the probabilities using the same principles of Bayes' theorem but including the new evidence.

The updated 'prior' probability is no longer the original 15%, but the 41% we calculated after hearing from the first witness. After running the numbers again, using Bayes' formula, the revised probability that the taxi was green increases to 86%.

INTERPRETING WITNESS TESTIMONIES WITH BAYES' THEOREM

Let's dive a bit deeper into the implications of these results. Here are some situations that may seem counterintuitive at first, but make sense when we apply Bayes' theorem:

1. **The 50-50 Witness:** Suppose we have a witness who is only right half the time—in other words, they are as likely to be right as they are to be wrong. Our intuition tells us that such a witness is adding no useful information, and Bayes' theorem agrees. The testimony of such a witness doesn't change our prior estimate.
2. **The Perfect Witness:** Now, imagine a witness who is always right—they have a 100% accuracy rate in identifying the taxi colour. In this case, if they say the taxi was green, then it must have been green. Bayes' theorem concurs with this conclusion.
3. **The Always-Wrong Witness:** What about a witness who always gets the colour wrong? In this case, if they say the taxi is green, then it must have been blue. Bayes' theorem agrees. We can trust this witness by assuming the opposite of what they say is the truth.

In summary, a 50% accurate witness adds no value to our estimate. A 100% accurate witness's testimony is definitive. An always-wrong witness inversely confirms the truth.

THE BASE RATE FALLACY AND ITS IMPLICATIONS

The Base Rate Fallacy occurs when we don't give enough weight to 'base rate' information (like the overall number of blue and green taxis) when making probability judgments. This mistake can lead us to overvalue specific evidence (like a single eyewitness account) and undervalue more general information like the ratio of blue to green taxis. Even so, the eyewitness may still be correct.

Again, if someone loves talking about books, we might intuitively guess that they are more likely to work in a bookstore or library than as, say, a nurse. But there are many more nurses than there are librarians or bookstore employees, and many of them love books. So, taking account of the base rate, we may well conclude that it's more likely that the book enthusiast is a nurse than a bookstore employee or librarian.

AVOIDING THE BASE RATE FALLACY

The Base Rate Fallacy leads us to ignore general information (like the ratio of blue to green taxis or nurses to librarians) in favour of specific evidence (an eyewitness account or specific bit of information). It's essential to balance specific and general information to avoid skewed judgments.

THE UNVEILING OF THE TRUTH

In the case of the New Brighton Taxi Problem, the mystery was solved when CCTV footage surfaced. The taxi involved was revealed to be yellow, a twist no one expected. Not really—there are no yellow taxis in New Brighton. In fact, both eyewitnesses were correct and the taxi was green.

CONCLUSION: TRUTH AND TESTIMONY

While our story was hypothetical, the principles it illustrates are very real and applicable in a wide variety of situations and circumstances. Bayes' theorem, base rates, and new evidence are all important parts of the detective's toolkit.

When Should We Believe the Diagnosis? Exploring the World of False Positives

THE FLU TEST SCENARIO: SETTING THE STAGE

Imagine this scenario: you twist your knee in a skateboarding mishap and decide to visit your doctor to have it looked at, just to be on the safe side. At the surgery, they run a routine test for a flu virus on all their patients, based on the estimate that about 1 out of every 100 patients visiting them will have the virus. This flu test is known to be pretty accurate—it gets the diagnosis right 99 out of 100 times. In other words, it correctly identifies 99% of people who are sick as sick, and equally importantly, it correctly clears 99% of those who don't have the flu virus.

Now, you take the test, and to your surprise, it comes back positive. What does this mean for you, exactly? You dropped in to have your knee looked at, and now it seems you have the flu.

To summarise the situation, imagine you've twisted your knee and, while at the doctor's office, you're given a routine flu test. The test is 99% accurate and is positive. But what are the actual chances that you have the flu? This scenario is perfect for exploring Bayes' theorem and understanding false positives.

BREAKING DOWN THE INVERSE FALLACY

Here, we step into the tricky territory of probabilities, a place where common sense can often mislead us. So, what is the chance that you do have the virus?

The intuitive answer is 99%, as the test is 99% accurate. But is that right?

The information we are given relates to the probability of testing positive given that you have the virus. What we want to know, however, is the probability of having the virus given that you test positive. This is a crucial difference.

Common intuition conflates these two probabilities, but they are very different. If the test is 99% accurate, this means that 99% of those with the virus test positive. But this is NOT the same thing as saying that 99% of patients who test positive have the virus. This is an example of the 'Inverse Fallacy' or 'Prosecutor's Fallacy'. In fact, those two probabilities can diverge markedly.

To summarise, common sense might suggest a 99% chance of having the flu, aligning with the test's accuracy. However, this confuses the probability of testing positive when having the flu with the probability of having the flu when testing positive—a common mistake known as the 'Inverse Fallacy'.

So what is the probability you have the virus if you test positive, given that the test is 99% accurate? To answer this, we can use Bayes' theorem.

APPLYING BAYES' THEOREM

Bayes' theorem, as we have seen, uses three values:

a. Your initial chance of having the flu before taking the test, which in our scenario was estimated to be 1 out of 100 or 0.01.
b. The likelihood of the test showing a positive result if you have the flu, which we know to be 99% or 0.99 based on the accuracy of the test.
c. The likelihood of the test showing a positive result if you don't have the flu, which is 1% or 0.01, again based on the accuracy of the test.

When we plug these into Bayesian formula, we end up with a surprising result. If you test positive for the flu, despite the test being 99% accurate, there's actually only a 50% chance that you really have it.

In other words, to find the real probability of having the flu, we consider:

1. **Prior Probability:** Your initial chance of having the flu is 1% (1 in 100).
2. **True Positive Rate:** The test correctly identifies the flu 99% of the time.
3. **False Positive Rate:** The test incorrectly indicates flu in healthy individuals 1% of the time.

The formula is expressed as follows:

$$ab/[ab+c\,(1-a)]$$

where

- a is the prior probability, i.e. 0.1,
- b is 0.99.
- c is 0.01.

Using Bayes' theorem, we find a surprising result: even with a 99% accurate test, there's only a 50% chance you have the flu after a positive result.

GRAPPLING WITH PROBABILITIES

The result can seem counterintuitive, and it's worth taking a moment to understand why that is. The key is to remember that the flu is a relatively rare occurrence—only 1 in 100 patients have it. While the test may be 99% accurate, we have to take into account the relative rarity of the disease in those who are tested. The chance is just 1 in 100. The chance of having the flu before taking the test and the chance of the test making an error are both, therefore, 1 in 100. These two probabilities are the same, and so, when you test positive, the chance that you have the flu is actually just 1 in 2.

It is basically a competition between how rare the virus is and how rarely the test is wrong. In this case, there is a 1 in 100 chance that you have the virus before taking the test, and the test is wrong one time in 100. These two probabilities are equal, so the chance that you have the virus when testing positive is 1 in 2, despite the test being 99% accurate.

Put another way, the counterintuitive outcome arises because the flu is relatively rare (1 in 100), balancing against the test's accuracy.

THE IMPLICATION OF SYMPTOMS AND PRIOR PROBABILITIES

This calculation changes if we add in some more information. Let's say you were already feeling unwell with flu-like symptoms before the test. In this case, your doctor might think you're more likely to have the flu than the average patient, and this would increase your 'prior probability'. Consequently, a positive test in this context would be more indicative of actually having the flu, as it aligns with both the symptoms and the test result.

In this way, Bayes' theorem incorporates both the statistical likelihood and real-world information. It's a powerful tool to help us understand probabilities better and to make informed decisions. The bottom line, though, is that while a positive test result can be misinterpreted, it should, especially in conjunction with symptoms, be taken seriously.

The Role of Symptoms in Adjusting Probabilities

If you had flu-like symptoms before the test, this would increase your 'prior probability'. Consequently, a positive test in this context would be more indicative of actually having the flu, as it aligns with both the symptoms and the test result.

CONCLUSION: THE BROAD APPLICATION OF BAYESIAN THINKING

While we've used the example of a flu test, the principles of Bayes' theorem apply beyond the doctor's door. From the courtroom to the boardroom, from deciding if an email is spam to weighing up the reliability of a rumour, we often need to update our beliefs in the face of new evidence. Remember, a single piece of evidence should always be weighed against the broader context and initial probabilities.

When Should We Believe That Something Is Rare? Exploring the Bayesian Beetle Problem

ENCOUNTERING A BEETLE: AN ADVENTURE IN PROBABILITY

Imagine you're on a wilderness walk and spot a beetle with an alluring pattern. This pattern is common (98%) on a rare type of beetle, which constitutes only 0.1% of the beetle population. It also appears on 5% of the common beetles. What's the likelihood that the beetle you found is one of the rare ones?

INTUITION VS. BAYESIAN LOGIC

Intuitively, seeing the rare pattern might make you think you've found a rare beetle. However, Bayes' theorem offers a more logical approach to evaluating the odds.

CALCULATING THE ODDS: A BAYESIAN APPROACH

1. **Initial Probability (Prior):** The rare beetles are only 0.1% of the population.
2. **Likelihood of Observing the Pattern on a Rare Beetle:** 98% of rare beetles have this pattern.
3. **Likelihood of Observing the Pattern on a Common Beetle:** 5% of common beetles have the pattern.

Despite the pattern being associated with the rare beetle, its rarity (0.1%) and the pattern's occurrence on common beetles (5%) significantly influence the probability.

THE PROBABILITY TWIST: UNRAVELLING THE PARADOX

Applying Bayes' theorem, we find that the probability of the beetle being rare, despite having the special pattern, is surprisingly low—only about 1.92%. This outcome may seem paradoxical, but it underscores how the rarity of the beetles and the pattern's presence on common beetles tilt the odds. The keyword here is 'rare'. These beetles are, by their very definition, few and far between. So, even with the special pattern, the odds are still in favour of you having stumbled upon one of the many common beetles.

UNDERSTANDING THE SIGNIFICANCE OF BASE RATES

The key takeaway is the significance of base rates in probability assessments. The rare beetle's low base rate (0.1%) dramatically affects the likelihood that any beetle with the pattern is rare. It's a vivid example of how base rates can challenge our intuition.

THE SUBTLE ART OF DISTINGUISHING PROBABILITIES

In summary, the chance that the beetle you're observing is rare, given that it sports the distinct pattern, is much less than the chance that a common beetle will have the pattern. This is because the prevalence of rare beetles is a huge order of magnitude less than that of common beetles. The base rate probability that you come across a rare beetle is just 0.1% or 0.001, while there's a much bigger chance that the beetle with the unusual pattern is common (5% or 0.05), Using Bayes' theorem, the chance that the beetle is rare is just 1.92%.

CONCLUSION: A RARE ENCOUNTER?

This encounter with the beetle, as trivial as it might seem, provides us with a key lesson in understanding probabilities and making informed judgments. It also illustrates an essential principle in probability and decision-making. Whether in natural observation, legal evidence, or everyday judgments, base rates are crucial. They shape our understanding of rarity and probability, guiding us to draw more informed and accurate conclusions.

When Should We Expect Success? Exploring the Bobby Smith Problem

INTRODUCTION: THE WORLD OF BOBBY SMITH, A BUDDING TENNIS PRODIGY

Bobby Smith, a young tennis player, faces daunting odds in his journey to professional status. In his world, 1 in 1,000 schoolboy tennis players make it to the professional ranks.

THE TEST: BOBBY'S GATEWAY TO THE ACADEMY

Bobby takes a crucial test to join the prestigious tennis academy, which serves as a breeding ground for future professionals. Though he passes the test and enters the academy, we must consider what this really indicates about his chances of turning pro.

THE FALLACY: MISINTERPRETING PROBABILITIES

It's crucial not to confuse the probability of Bobby turning pro (given his academy entry) with the inverse—the probability of him entering the academy if destined to turn pro. While all professionals come from the academy, not all academy members become professionals.

THE GATEKEEPER: A SPECIAL TEST

Bobby is given a test designed to gauge the potential of young tennis players, which is used to determine who will have the privilege of becoming a member of the tennis academy, a training and nurturing ground for aspiring professionals. Bobby takes this test with the goal of securing membership.

THE CHALLENGE: OVERCOMING THE ODDS

The test is taken by a thousand of these budding tennis players, including Bobby, all of whom want to enter the academy. Just 5% of those tested will gain entry to the academy and then fail to become professional players. One will succeed. In other words, of the 1,000 people who take the test, 50 will enter the academy but not turn pro, while only one will succeed.

Graduation from the academy is also a condition of entry to the professional tour in Bobby's world. As such, we can rule out anyone who does not gain entry to the academy as a future professional player.

BOBBY'S TRIUMPH: ENTERING THE ACADEMY

Fortunately for Bobby, he aces the test and joins the academy. This is a crucial step for him. After all, every professional player in Bobby's world, as we have noted, is a graduate of the academy.

It might now seem almost certain that he has a bright future ahead in the world of professional tennis. But is this a correct assessment?

Well, it is undeniable that without entrance to the academy there is no way for Bobby to achieve professional status, but he has aced the test and is now a member of the academy. Give the accuracy of the test in sifting talent, can we now look forward with some confidence to his future sporting career?

Well, determining the probability of Bobby becoming a professional tennis player if he scores well enough on the test to gain entry to the academy is a complex matter. It involves factors beyond just his entrance to the academy. Many other elements, such as his dedication, talent, and the competitive environment, play roles in determining his chances. Even so, it does look promising, or does it?

THE FALLACY: AN ILLUSION OF CERTAINTY

Back to the test result, we must be very careful not to confuse the probability of Bobby going on to a professional tennis career given his entrance to the academy with its inverse—the probability that he would enter the academy if he were to go on to attain the professional ranks.

In our example, the probability of his entrance to the academy given that Bobby will make it to professional circles is a sure thing. All future professional players will be graduates of the academy. What we seek to know, however, is something different—it is the probability that Bobby will become a professional player given that he enters the academy. This is a very different question.

Put another way, the fallacy arises from confusing two distinct probabilities:

1. The probability of a hypothesis being true (Bobby will become a professional tennis player) given some evidence (entrance to the academy).
2. The probability of the evidence (entrance to the academy) given the hypothesis is true (Bobby will become a professional player).

In simple terms, if we know that Bobby became a professional, he definitely went to the academy. But that's not what we're interested in. We want to know the odds of Bobby becoming a professional, given that he got into the academy.

So what is the actual chance that Bobby will become a professional tennis player if he scores well enough on the test to gain entry to the academy?

CALCULATING THE REAL PROBABILITY: BEWARE OF FALSE POSITIVES

When we dig deeper into the data, we uncover some revealing insights. Consider the 5% of students who pass the test and enter the academy but don't go on to become professional players—they are the 'false positives' in our scenario. If we assume 1,000 students take the test, 50 such 'false positives' get into the academy.

Add to them the one student who does become a pro (from the original pool of 1,000), and you find that Bobby's chances of turning pro, even after making it into the academy, are just 1 in 51. This translates to approximately 1.96%.

This will only change if we know some additional information about Bobby.

THE MEDICAL ANALOGY: VIRUS TESTING

Interestingly, this concept aligns with the 'false positives' problem in the medical field, particularly in regard to virus testing. Let's take a group of 1,000 people getting tested for a certain virus. Even with a test accuracy of 95%, about 5% of those tested (50 individuals) will also test positive despite not carrying the virus. On top of these, one individual does have the virus. Thus, if you test positive, the probability of actually carrying the virus is again about 1.96%, unless there is some additional information we need to take into account.

A MATHEMATICAL ASSURANCE: BAYES' THEOREM

Though we've already figured out Bobby's chances of turning pro, there's another way to confirm our findings. This alternative method involves Bayes' theorem. This theorem helps us calculate the updated probability of a hypothesis (in our case, Bobby turning pro) after obtaining new evidence (Bobby entering the academy).

The formula is expressed as follows:

$ab/[ab+c\,(1-a)]$

where

- *a* is the prior probability, i.e. the probability that Bobby will turn pro before we know his test results (0.001, as Bobby is one among 1,000),
- *b* is the probability of Bobby entering the academy if he will turn pro (which is 100%, as all pros in Bobby's world are academy graduates), and
- *c* is the probability of Bobby entering the academy if he won't turn pro (which is 50 out of 999, as out of the 999 kids who won't turn pro, 50 will enter the academy).

By plugging these values into Bayes' theorem, we confirm that Bobby's chances of becoming a professional, despite gaining entry to the academy, are not 95% as one might think, but around 1.96%.

CRUNCHING NUMBERS: THE HARD REALITY

To summarise, let's analyse the situation numerically. Among the 1,000 kids applying for the academy, 50 will be accepted but won't make it to professional status. One will eventually turn pro. So, out of the 51 kids admitted, only one will become a professional. Therefore, the chance of becoming a professional tennis player if you enter the tennis academy is 1 in 51, or roughly 1.96%, unless there is some additional information that we need to take into account.

THE TWIST: A SUCCESS STORY

Despite the low probability, Bobby turns out to be the exception. He defies the odds and ends up winning the Australian Open under a different name.

CONCLUSION: BEYOND THE NUMBERS

Bobby's story highlights how statistical probabilities can mislead our intuition. Understanding these concepts is crucial, whether assessing the future of a tennis player or interpreting medical test results. Despite the odds, individuals like Bobby can defy statistics, reminding us that while numbers describe populations, they don't predetermine individual destinies.

When Should We Close the Case? Exploring the Kingfisher Manor Mystery

DISCOVERING THE BODY

Set on the brooding Moors of Southwest England stands Kingfisher Manor, a stately and imposing presence with a storied past. Its hallways and long corridors tell tales of intrigue and menace, and its ancient walls hold secrets that have never been exposed. This is a tale of one fateful summer's afternoon, on which the manor's eerie silence was shattered by the discovery of Lord Montgomery-Newton, a renowned archaeologist known for his documentation of the secrets of the Egyptian Pharaohs. He lay lifeless on the floor, grasping to his chest an ancient Egyptian amulet, believed by some to hold mystical and medical powers.

News was quick to spread of the tragic fate of the noted academic and within the hour Detective Inspector Anna Marchbank was at the scene, ready and eager to unravel the mystery that lay before her.

UNVEILING THE SUSPECTS

As the chaos resulting from the initial shock died down, DI Marchbank looked around the room, scanning every detail. The study was home to elegant Edwardian furniture and bookshelves filled with rare volumes and manuscripts. It was already clear that Lord Montgomery-Newton's death was no accident—the discarded revolver was ample testimony to that.

Turning her attention to the suspects, Marchbank learned that five individuals had been present at the manor on the day of the murder. Each of these exuded some degree of suspicion, their personal secrets intertwining with those of the ancient abode.

1. **Mr. Hadleigh:** A loyal servant of the household for over 30 years, Mr. Hadleigh was known for his profound devotion and loyalty towards Montgomery-Newton. However, there were rumours of a grudge stemming from a well-guarded incident in the distant past.
2. **Captain Blackwood:** A career soldier with a charming side, Captain Blackwood had recently returned from a secret military operation overseas. Whispers of his involvement in clandestine affairs had long circulated.

3. **Dr. Winterbottom:** A well-respected doctor with a fascination for the history of medicine, Dr. Winterbottom's special interest in the ancient dark arts made her an intriguing suspect.
4. **Miss Sinclair:** The beautiful and enigmatic Miss Sinclair was a frequent guest at Kingfisher Manor, although her precise connection to the Montgomery-Newton family was veiled in mystery. The motivation for her ubiquitous presence had raised suspicions in the eyes of some.
5. **Professor Adamant:** A distant relative of Mr. Hadleigh, Professor Adamant, an expert in medieval theology, had arrived at the manor just days before the murder. His debonair bearing and Edwardian dress sense marked him out from the crowd.

With the suspects identified, Detective Inspector Marchbank knew that putting together the pieces of this jigsaw would require a sharp mind and a keen focus. She had enough information to know that there was only one person involved in the murder of the noble Lord and started by assigning an equal probability to each of the suspects, assigning them each a 20% chance of being the guilty party.

THE DANCE OF CLUES

The investigation commenced, and DI Marchbank began the process of putting together the fragments of evidence.

TWO HOURS INTO THE INVESTIGATION: ELIMINATING MR. HADLEIGH

The first breakthrough came when an airtight alibi emerged for Mr. Hadleigh. He had been attending a high-profile charity event in the neighbouring village at the time of the murder. The detective swiftly eliminated him from the list of suspects, narrowing down the field to Captain Blackwood, Dr. Winterbottom, Miss Sinclair, and Professor Adamant.

With Mr. Hadleigh's elimination, the probability of guilt for the remaining four suspects increased to 25% each. Marchbank knew that each subsequent clue would alter these probabilities, moving her closer to the elusive killer.

FOUR HOURS IN: QUESTIONING CAPTAIN BLACKWOOD'S ALIBI

As the investigation progressed, doubts began to emerge regarding Captain Blackwood's alibi. Witness testimonies conflicted, creating a fog of uncertainty around his whereabouts on the afternoon of the murder. Marchbank sensed a crack in his armour and increased her evaluation of his probability of guilt to 40%.

The detective understood the delicate balance of probabilities, acknowledging the importance of assigning weight to each suspect based on the available evidence. She embraced the Bayesian approach, allowing it to guide her through the labyrinthine twists of the investigation.

SIX HOURS AND 45 MINUTES IN: EXONERATING DR. WINTERBOTTOM

The Detective Inspector was made aware of a crucial piece of evidence that placed Dr. Winterbottom far from the scene of the crime. Reliable witnesses confirmed her presence at a medical conference during the time of the murder, eliminating her as a suspect.

The investigation was gaining momentum, but the truth still eluded them. Marchbank looked for the thread that could potentially unravel the entire tapestry of deception.

THE BAYESIAN BALANCE

As Detective Inspector Marchbank meticulously evaluated the evidence, she was acutely aware that each new piece of evidence needed to be filtered through the prism of prior probabilities.

Analysing the case through this Bayesian lens, she considered the individual probabilities assigned to each suspect. Captain Blackwood had been assigned a probability of guilt of 40%, leaving a 60% chance of it being one of the other remaining suspects. As such, she was able to assign an equal 20% probability of it being Miss Sinclair, Professor Adamant, or Dr. Winterbottom. Now that Winterbottom had been eliminated, her 20% share of the probability needed to be distributed to the other suspects. Critically, the Bayesian approach dictated that Captain Blackwood's probability of guilt should be adjusted twice as much as the probabilities for the other two suspects in this process, since his prevailing assigned chance of being the culprit (40%) was twice that of each of the others (Miss Sinclair and Professor Adamant) before the Doctor was eliminated.

Marchbank was quick to raise Captain Blackwood's probability of guilt, therefore, by 10%, to 50%, reflecting the weight of the evidence against him. Simultaneously, she increased the probabilities assigned to Miss Sinclair and Professor Adamant from 20% to 25% each.

SUMMARISING THE INVESTIGATION

To summarise where we have got to, Marchbank identified five suspects, each with potential motives and secrets. Initially, she assigned each a 20% probability of guilt, and then used a Bayesian approach to adjust these probabilities as new evidence emerges.

THE INVESTIGATION: A SERIES OF BREAKTHROUGHS

- **Eliminating Mr. Hadleigh:** An airtight alibi for Mr. Hadleigh removes him from suspicion, increasing the remaining suspects' probabilities to 25% each.
- **Doubting Captain Blackwood's Alibi:** Conflicting testimonies about Captain Blackwood raise his probability of guilt to 40%.
- **Exonerating Dr. Winterbottom:** Evidence places Dr. Winterbottom away from the scene, eliminating her as a suspect.

REASSESSING PROBABILITIES

With each clue, Marchbank recalculates the probabilities:

- **Captain Blackwood's Increased Chances:** After eliminating Dr. Winterbottom, Blackwood's probability of guilt rises to 50%, reflecting the growing suspicion against him.
- **Remaining Suspects:** Miss Sinclair and Professor Adamant's probabilities increase to 25% each.

THE TRIAL AND TWISTS OF FATE

The revised probabilities paved the way for the trial of Captain Blackwood, his fate hanging in the balance.

Trapped within the walls of a courtroom, Blackwood was at the mercy of a prosecution barrister who skilfully guided the jury to the trap door of the Prosecutor's Fallacy. Like so many juries before them, they confused the likelihood that someone is guilty in light of the evidence with the likelihood of observing the evidence if they were guilty. The likelihood that Montgomery-Newton was killed in the study if the Captain was guilty of his murder was naturally rather high, and this led to his conviction. Unfortunately for Captain Blackwood, the relevant probability (that he was guilty of murder given that the great man was killed in the study) was somewhat smaller but bypassed in the deliberations.

It's certainly true that the evidence was consistent with the Captain's guilt. Yet it was equally consistent with the guilt of the other suspects. But they were not in the dock! Unfortunately for the Captain, he was. The verdict of guilty was not long in coming.

The Trial and Error: The Prosecutor's Fallacy

The jury had fallen prey to the Prosecutor's Fallacy, confusing the likelihood of Captain Blackwood's guilt given the evidence with the likelihood of the evidence if he were guilty. This led to his wrongful conviction.

The Twist

The true killer, Miss Sinclair, evaded justice entirely. Concealed within the pages of an ancient manuscript, she had hidden a letter, a damning piece of evidence linking her to a nefarious smuggling operation. Lord Montgomery-Newton's discovery of the letter sealed his fate. In a desperate act to protect her secrets, Miss Sinclair had resorted to murder.

Captain Blackwood, a victim of circumstance and statistical misinterpretation, would serve a life sentence for a crime he did not commit. Meanwhile, Miss Sinclair eluded the clutches of justice, disappearing into the shadows as a tax exile in a distant land.

CONCLUSION: THE COMPLEXITY OF TRUTH AND PROBABILITY

Kingfisher Manor, once a place of mystery and intrigue, stands now as a solemn witness to the twists and turns of fate. The echoes of the past whisper through its halls, reminding us of the delicate balance between evidence and probability. More broadly, the Kingfisher Manor mystery underscores the intricacies of evidence interpretation and probability assessment. It highlights the challenges in drawing conclusions from circumstantial evidence and emphasises the critical role of Bayesian reasoning in great investigative endeavours.

When Should We Trust a Loved One?
Exploring a Shakespearean Tragedy

OTHELLO: THE BACKGROUND

Created by William Shakespeare, 'Othello' is a play centred around four main characters: Othello, a general in the Venetian army; his devoted wife, Desdemona; his trusted lieutenant, Cassio; and his manipulative ensign, Iago. Iago's plan forms the central conflict of the play. Driven by jealousy and a large helping of evil, Iago seeks to convince Othello that Desdemona is conducting a secret affair with Cassio. His strategy hinges on a treasured keepsake, a precious handkerchief which Desdemona received as a gift from Othello. Iago conspires successfully to plant this keepsake in Cassio's lodgings so that Othello will later find it.

UNDERSTANDING OTHELLO'S MINDSET

Othello's reaction to this discovery can potentially take different paths, depending on his character and mindset. If Othello refuses to entertain any possibility that Desdemona is being unfaithful to him, then no amount of evidence could ever change that belief.

On the other hand, Othello might accept that there is a possibility, however small, that Desdemona is being unfaithful to him. This would mean that there might be some level of evidence, however overwhelming it may need to be, that could undermine his faith in Desdemona's loyalty.

There is, however, another path that Othello could take, which is to evaluate the circumstances objectively and analytically, weighing the evidence. But this balanced approach also has its pitfalls. A very simple starting assumption that he could make would be to assume that the likelihood of her guilt is equal to the likelihood of her innocence. That would mean assigning an implicit 50% chance that Desdemona had been unfaithful. This is known as the 'Prior Indifference Fallacy'. If the prior probability is 50%, this needs to be established by a process better than simply assuming that because there are two possibilities (guilty or innocent), we can ascribe automatic equal weight to each. If Othello falls into this trap, any evidence against Desdemona starts to become very damning.

THE LOGICAL CONTRADICTION APPROACH

An alternative approach would be to seek evidence that directly contradicts the hypothesis of Desdemona's guilt. If Othello could find proof that logically undermines the idea of her infidelity, he would have a solid base to stand on. However, there is no such clear-cut evidence, leading Othello deeper into a mindset of anger and suspicion.

BAYES' THEOREM TO THE RESCUE

Othello might seek a strategy that allows him to combine his subjective belief with the new evidence to form a rational judgement. This is where Bayes' theorem comes in. Bayes' theorem allows, as we have seen in previous chapters, for the updating of probabilities based on observed evidence. The theorem can be expressed in the following formula:

Updated probability $= ab/[ab+c\,(1-a)]$

In this formula, a is the prior probability, representing the likelihood that a hypothesis is true before encountering new evidence. b is the conditional probability, describing the likelihood of observing the new evidence if the hypothesis is true. And finally, c is the probability of observing the new evidence if the hypothesis is false. In this case, the evidence is the keepsake in Cassio's lodgings, and the hypothesis is that Desdemona is being unfaithful to Othello.

APPLYING BAYES' THEOREM
TO OTHELLO'S DILEMMA

Now, before he discovers the keepsake (new evidence), suppose Othello perceives a 4% chance of Desdemona's infidelity ($a=0.04$). This represents his prior belief, based on his understanding of Desdemona's character and their relationship. Of course, he is not literally assigning percentages, but he is doing so implicitly, and here we are simply making these explicit to show what might be happening within a Bayesian framework.

Next, consider the probability of finding the keepsake in Cassio's room if Desdemona is indeed having an affair. Let's assume that Othello considers there is a 50% chance of this being the case ($b=0.5$).

Finally, what is the chance of finding the keepsake in Cassio's room if Desdemona is innocent? This would in Othello's mind require an unlikely series of events, such as the handkerchief being stolen or misplaced, and then ending up in Cassio's possession. Let's say he assigns this a low probability of just 5% ($c = 0.05$).

BAYESIAN PROBABILITIES: WEIGHING THE EVIDENCE

Feeding these values into Bayes' equation, we can calculate the updated (or posterior) probability of Desdemona's guilt in Othello's eyes, given the discovery of the keepsake. The resulting probability comes out to be 0.294 or 29.4%. This suggests that, after considering the new evidence, Othello might reasonably believe that there is nearly a 30% chance that Desdemona is being unfaithful.

IAGO'S MANIPULATION OF PROBABILITIES

This 30% likelihood might not be high enough for Iago's deceitful purposes. To enhance his plot, Iago needs to convince Othello to revise his estimate of c downwards, arguing that the keepsake's presence in Cassio's room is a near-certain indication of guilt. If Othello lowers his estimate of c from 0.05 to 0.01, the revised Bayesian probability shoots up to 67.6%. This change dramatically amplifies the perceived impact of the evidence, making Desdemona's guilt appear significantly more probable.

DESDEMONA'S DEFENCE STRATEGY

On the other hand, Desdemona's strategy for defending herself could be to challenge Othello's assumption about b. She could argue that it would be illogical for her to risk the discovery of the keepsake if she were truly having an affair with Cassio. By reducing Othello's estimate of b, she can turn the tables and make the presence of the keepsake testimony to her innocence rather than guilt.

CONCLUSION: THE TIMELESS BAYESIAN

Shakespeare's 'Othello' was written about a century before Thomas Bayes was born. Yet the complex interplay of trust, deception, and evidence in the tragedy presents a classic case study in Bayesian reasoning.

Shakespeare was inherently Bayesian in his thinking. The tragedy of the play is that Othello was not!

When Should We Trust the Jury?
Exploring a Courtroom Tragedy

THE CONVICTION

In the final weeks of the 20th century, a lawyer named Sally Clark was convicted of the murder of her two infant sons. Despite being a woman of good standing with no history of violent behaviour, Clark was swept up in a whirlwind of accusations, trials, and appeals that would besmirch the criminal justice system and cost her dearly.

THE INVESTIGATION AND TRIAL—
BUILDING A CASE ON UNCERTAINTY

The deaths of Clark's two children were initially assumed to be tragic instances of Sudden Infant Death Syndrome (SIDS), a cause of infant mortality that was not well understood even by medical experts. However, the authorities became suspicious of the coincidental deaths, leading to Clark's eventual trial. As the investigation evolved, it subsequently transpired that numerous pieces of evidence helpful to the defence were withheld from them.

STATISTICAL EVIDENCE—THE MISINTERPRETATION

The prosecution presented a piece of seemingly damning statistical evidence during Clark's trial. One of their witnesses, a paediatrician, asserted that the probability of two infants from the same family dying from SIDS was incredibly low—approximately 1 in 73 million. He compared the odds to winning a bet on a longshot in the iconic Grand National horse race four years in a row.

THE PROSECUTOR'S FALLACY—THE DANGEROUS CONFLATION OF PROBABILITIES

The flaws in the statistical argument presented at the trial were both substantial and consequential. The paediatrician had mistakenly assumed that the deaths of Clark's children were unrelated, or 'independent' events. This assumption neglects the potential for an underlying familial or genetic factor that might contribute to SIDS.

Moreover, the paediatrician's argument represents a common misinterpretation of probability known as the 'Prosecutor's Fallacy'. This fallacy involves conflating the probability of observing specific evidence if a hypothesis is true, with the probability that the hypothesis is true given that evidence. These are two very different things but easy for a jury of laymen to confuse.

THE PROSECUTOR'S FALLACY EXPLAINED

This fallacy arises from confusing two different probabilities:

1. The probability of observing specific evidence (in this case, two SIDS deaths) if a hypothesis (Clark's guilt) is true.
2. The probability that the hypothesis is true given the observed evidence.

THE NEED FOR COMPARATIVE LIKELIHOOD ASSESSMENT

The Royal Statistical Society emphasised the need to compare the likelihood of the deaths under each hypothesis—SIDS or murder. The rarity of two SIDS deaths alone doesn't provide sufficient grounds for a murder conviction.

PRIOR PROBABILITY—UNDERSTANDING THE LIKELIHOOD OF GUILT BEFORE THE EVIDENCE

Prior probability—a concept integral to understanding the Prosecutor's Fallacy—is often overlooked in court proceedings. This term refers to the probability of a hypothesis (in this case, that Sally Clark is a child killer) being true before any evidence is presented.

Given that she had no history of violence or harm towards her children, or anyone else, or any indication of such a tendency, the prior probability of her being a murderer would be extremely low. In fact, the occurrence of two cases of SIDS in a single family is much more common than a mother murdering her two children.

The jury should weigh up the relative likelihood of the two competing explanations for the deaths. Which is more likely? Double infant murder by a mother or double SIDS?

More generally, it is likely in any large enough population that one or more cases of something highly improbable will occur in any particular case.

In a letter from the President of the Royal Statistical Society to the Lord Chancellor, Professor Peter Green explained the issue succinctly:

> The jury needs to weigh up two competing explanations for the babies' deaths: SIDS or murder. The fact that two deaths by SIDS is quite unlikely is, taken alone, of little value. Two deaths by murder may well be even more unlikely. What matters is the relative likelihood of the deaths under each explanation, not just how unlikely they are under one explanation.

Put another way, before considering the evidence, the prior probability of Clark being a murderer, given her background and lack of violent history, was extremely low. The probability of two SIDS deaths in one family, while rare, was still much higher than the likelihood of the mother murdering her two children.

THE NEED FOR COMPARATIVE LIKELIHOOD ASSESSMENT

The Royal Statistical Society emphasised the need to compare the likelihood of the deaths under each hypothesis—SIDS or murder. The rarity of two SIDS deaths alone doesn't provide sufficient grounds for a murder conviction.

The Fictional Case of Lottie Jones

To illustrate the Prosecutor's Fallacy, consider the fictional case of Lottie Jones, charged with winning the lottery by cheating. The fallacy occurs when the expert witness equates the low probability of winning the lottery (1 in 45 million) with the probability that a lottery win was achieved unfairly.

As in the Sally Clark case, the prosecution witness in this fictional parody commits the classic 'Prosecutor's Fallacy'. He assumes that the probability Lottie is innocent of cheating, given that she won the Lottery, is the same thing as the probability of her winning the Lottery if she is innocent of cheating. The former probability is astronomically higher than the latter unless we have some other indication that Lottie has cheated to win the Lottery. It is a clear example of how it is likely, in any large enough population, that things will happen that are improbable in any particular case. In other words, the 1 in 45 million represents the probability that a Lottery entry at random will win the jackpot, not the probability that a player who has won did so fairly!

Lottie just got very, very lucky just as Sally Clark got very, very unlucky.

THE AFTERMATH—TRAGEDY AND LESSONS LEARNED

Following her acquittal in 2003, Sally Clark never recovered from her ordeal and sadly died just a few years later. Her story stands as testament to the potential for disastrous consequences when statistics are misunderstood or misrepresented.

O.J. SIMPSON—AN ALTERNATE SCENARIO

Even in high-profile cases, such as American former actor and NFL football star O.J. Simpson's murder trial in the 1990s, this same misinterpretation of statistics is prevalent. Simpson's defence team argued that it was unlikely Simpson killed his wife because only a small percentage of spousal abuse cases result in the spouse's death. This argument, though statistically accurate, overlooks the relevant information—the fact that about 1 in 3 murdered women were killed by a spouse or partner. This represents a very clear case of the misuse of the Inverse or Prosecutor's Fallacy in argumentation before a jury.

CONCLUSION: THE IMPORTANCE OF STATISTICAL LITERACY

The importance of statistics in our justice system cannot be overstated. We must recognise the potential for misinterpretation and the potentially devastating results. A concerted effort to promote statistical literacy, particularly within our legal systems, can hopefully go a long way in preventing future miscarriages of justice.

A Question of Paradox

2

When Should We Change Our Mind? Exploring the Monty Hall Problem

THE GENESIS OF THE MONTY HALL PROBLEM

The Monty Hall Problem was named after the original host of the American game show, 'Let's Make a Deal'. It became a topic of popular debate because of the answer provided to a question quoted in a column in Parade magazine.

The concept is that contestants are given a choice of three doors. Behind one door lies a highly desirable prize like a car, while behind the other two doors were much less desirable prizes like goats. The car is placed randomly behind one of the doors, preventing contestants from predicting its location based on prior observations or information.

THE PUZZLE UNVEILED

Imagine yourself on this game show. You are asked to choose one of three doors (let's call them Doors 1, 2, and 3). After making your choice (let's say you choose Door 1), the host, who knows what's behind each door, opens another door (for instance, Door 3) to reveal a goat.

He then offers you a choice. You can stick with your original decision (Door 1 in this case), or you can switch to the remaining unopened door (Door 2). You should note that the host always opens a door that you didn't choose and that hides a goat, increasing the suspense and making the game more interesting.

The question that the Monty Hall problem asks is: Should you stick with your original choice, or should you switch to the other unopened door?

DOI: 10.1201/9781003402862-2

THE COUNTERINTUITIVE ANSWER

At first glance, it might seem like your odds of winning the car are the same whether you stick to your original choice or switch. After all, there are only two doors left unopened, so isn't there a 50% chance that the car is behind each of them?

In her column, Marilyn Vos Savant argued that the chance is not 50% either way, but that you have a higher chance of winning the car if you decide to switch doors. Despite receiving numerous objections from readers, including some leading academics, her answer holds up under scrutiny. Here's why.

When you first choose a door, there is a 1 in 3 chance that it hides the car. This means that there's a 2 in 3 chance that the car is behind one of the other two doors. Even after the host opens a door to reveal a goat, these probabilities do not change. Monty is simply providing more information about where the car is not.

So, if you stick with your original choice, your chances of winning the car remain at 1 in 3. However, if you switch, your chances increase to 2 in 3. Switching doors effectively allows you to select both of the other doors, doubling your odds of finding the car.

A CLOSER LOOK AT THE PROBABILITIES

Let's examine the situation more closely to understand how this works.

1. If the car is behind Door 1, and you choose it and stick with your choice, you win the car. The chance of this happening is 1/3.
2. If the car is behind Door 2 and you initially choose Door 1, the host will open Door 3 (since it conceals a goat). If you switch to Door 2, you win the car. The chance of this happening is 2/3.
3. If the car is behind Door 3, and you initially choose Door 1, the host will open Door 2 (since it conceals a goat). If you switch to Door 3, you win the car. The chance of this happening is also 2/3.

From the above, you can see that you have a 2/3 chance of winning if you switch to whichever door Monty has not opened, and a 1/3 chance of winning if you stick to your initial choice.

THE ROLE OF THE HOST

It's crucial to note that the host's knowledge and actions play a pivotal role in these probabilities. If the host didn't know what was behind each door or randomly chose a door to open, then the odds would indeed be 50–50, as he might have inadvertently opened a door to reveal the car. However, because the host always opens a door you didn't choose and always reveals a goat, the odds shift in favour of switching doors.

To expand upon this, consider a version of the problem with 52 cards. This time, you're invited to choose one card from a deck of 52. The objective is to select the Ace of Spades from a deck of cards lying face down on the table.

If you initially choose the Ace of Spades and stick with your choice, you win the game. The chance of this happening is 1/52, since there's only one Ace of Spades in a 52-card deck.

However, if you initially choose any card other than the Ace of Spades (which has a 51/52 chance), the host, knowing where the Ace of Spades is, will begin to turn cards over one at a time, always leaving the Ace of Spades and your initial card choice in the remaining face-down deck. The host will continue to do this until only your card and one other card remain. One of these two cards will be the Ace of Spades.

At this point, there is still a 1/52 chance that your original card is the Ace of Spades. If you switch your choice to the remaining card, the chance that it will be the Ace of Spades is therefore 51/52, which is a much higher probability than if you stick with your initial choice.

This works because the host each time deliberately turns over a card that is not the Ace of Spades. So the other card left face down at the end is either the Ace of Spades, with a chance of 51/52, or else your original choice is the Ace of Spades, with a probability of 1/52.

If the host doesn't know where the Ace of Spades is located, he might inadvertently reveal it every time he turns a card over, so he would be providing no new information about the location of the Ace of Spades by exposing a card.

This shows how the Monty Hall problem can scale to larger numbers. The initial odds of choosing the Ace of Spades are 1/52, but if you switch your choice after the host takes away all but one of the other cards, your odds improve dramatically to 51/52. This is a counterintuitive result, but it follows from the fact that the host's actions (because he knows where the Ace of Spades is) give you additional information about where the Ace of Spades is not.

OVERCOMING INTUITION WITH LOGIC

The Monty Hall problem can be difficult to grasp because it seems to contradict our intuition. The human brain tends to simplify complex situations, and when there are two unopened doors, it's easy to fall into the trap of assuming there's a 50% chance of winning either way. However, the Monty Hall problem highlights how understanding probability requires careful thought and a logical analysis of the situation.

EXPLORING THE MONTY HALL PROBLEM WITH SIMULATIONS

If you're still having trouble grasping the Monty Hall problem, you might find it helpful to see it in action. Numerous online simulators let you play the Monty Hall game repeatedly, and over time, you'll see that switching doors indeed wins about 2/3 of the time.

THE MONTY HALL PROBLEM IN POPULAR CULTURE

The Monty Hall problem has seeped into popular culture, appearing in films, television series, and even songs. It serves as a reminder that intuition and probability sometimes have a complicated relationship. The logical and statistical reasoning involved in this puzzle, as well as its seemingly paradoxical result, have made it a favourite topic in probability and statistics classes across the world.

CONCLUSION: PROBABILITY AND INTUITION

The Monty Hall problem is a captivating illustration of how probability can sometimes be counterintuitive. Although it's been debated, analysed, and confirmed many times over, it continues to intrigue and perplex. It provides a clear lesson: intuition isn't always reliable when it comes to probability.

When Should We Switch to Survive? Exploring the Deadly Doors Dilemma

THE GAME OF DESTINY: CHOOSING BETWEEN FOUR DEADLY DOORS

Welcome to a dark game of chance and destiny, featuring four distinctive doors coloured red, yellow, blue, and green. Three of these gateways lead to an instant, dusty demise, while the remaining one offers a golden path to fame and fortune. The destiny of each door is randomly assigned by the host, who picks out four coloured balls from a bag—red, yellow, blue, and green. This random process determines the fate that each door offers.

THE INITIAL CHOICE AND ODDS OF SURVIVAL

Suppose you find yourself drawn to the red door. Given the game's rules, your chance of picking the lucky door and moving onto a path of wealth and glory stand at just one in four, or 25%. Conversely, the unnerving possibility of your choice leading to a dusty doom looms large, with a daunting chance of three in four, or 75%. This calculation comes directly from the fact that out of the four doors, only one leads to fortune, while the other three lead to an unwelcome demise.

A TWIST IN THE TALE: THE HOST'S REVEAL

But the game involves a twist: the host, who knows where each door leads, opens one of the remaining doors. In this case, he reveals the yellow door to be one of the deadly ones. This is a part of the game's rule—the host must open a door after the initial choice, revealing one of the deadly doors while leaving the lucky door unopened.

THE PIVOTAL DECISION: TO SWITCH OR NOT TO SWITCH

With one door opened and its deadly fate exposed, you face a critical decision. Would you stick with your original choice, the red door, or change your fate by choosing either the blue or green door? This predicament is an extension of the classic three-door Monty Hall Problem, which we can term 'Monty Hall Plus', but the underlying logic is exactly the same.

THE COMMON MISCONCEPTION: MISUNDERSTANDING PROBABILITIES

Intuition might suggest that with one door less in the equation, the chance of the red door leading to fortune must have improved. After all, now there are only three doors left—the red, blue, and green. If we assume each door is now equally likely to be the lucky one, the probability of each would be one in three.

ANOTHER REVEAL, ANOTHER DEATH TRAP

However, the host has yet to finish his part. He proceeds to open another door, unveiling the blue one this time, which again turns out to be a death trap. Now, with only two doors remaining—the red and green—the odds seem to have further improved, right? The likelihood of each door leading to fortune should now stand at a clear 50-50, or does it? Does it matter if you stick with your original choice or switch to the remaining door?

THE COUNTERINTUITIVE TRUTH: WHY THE INITIAL CHOICE MATTERS

Contrary to intuitive reasoning, the answer is a resounding yes; it does matter if you stick or switch. The reason for this lies in the fact that the host knows what lies behind each door. When you initially chose the red door, your odds of it leading to fame and fortune were 25%. These odds remain unchanged if you persist with your original choice, regardless of which doors the host reveals subsequently.

THE VALUE OF INFORMATION: HOW THE HOST'S ACTIONS ALTER THE ODDS

Here lies the crux of the game—the host's actions, since they are informed, change the probabilities associated with the remaining doors. Before the host opened the yellow door, there was a 75% chance that the fortunate door was one among the yellow, blue, or green doors. But now, with the yellow door revealed as deadly, the same 75% probability now gets distributed to the remaining doors, i.e. the remaining (blue and green) doors.

THE FINAL REVEAL:
GREEN—THE FINAL OPTION

As the host opens the blue door, unveiling yet another deadly fate, the odds shift again. The chance of the green door being the fortunate one grows further, given that it is now the only door standing against your initial choice, the red door. Therefore, you could either stick with your original choice and hold onto the 25% chance of survival, or switch to the green door, enhancing your odds to a favourable 75%. Essentially, the combined probability of the doors not initially chosen (which was originally 3/4) now heavily favours the last unopened door (since two of three potential safe doors have been eliminated).

CONCLUSION: THE IMPLICATION
OF KNOWLEDGE

This dynamic interplay of choices and probabilities is a result of the host's knowledge about what lies behind each door. The host's actions introduce new information into the game and influence the probability associated with the remaining unopened doors. The odds change because the host, knowing the outcomes, will never inadvertently reveal the lucky door. However, if the host didn't possess this knowledge and the doors were revealed randomly, the game would lose its strategic aspect and boil down to sheer luck. In such a scenario, if two doors remain, the chances would be a clear 50-50, making a coin toss as effective a decision-making tool as any others.

When Should We Open the Casket?
Exploring the Suitors' Dilemma

THE THREE CASKET CONUNDRUM

The narrative of William Shakespeare's 'Merchant of Venice' contains intrigue around the character of the young heiress Portia. Amid the various plot developments, one of the more fascinating elements of the story lies in a puzzle set for anyone seeking her hand in marriage. Three caskets made of gold, silver, and lead each contain a different item. Only one holds the prize, a miniature portrait of Portia which symbolises the route to her heart. Portia alone knows that the portrait's true location is in the lead casket.

SUITORS AND THE CRYPTIC CASKETS: UNRAVELLING THE PECULIAR TEST

As the story unfolds, we learn that to claim Portia in holy wedlock a suitor must choose the casket that houses her portrait. Each casket comes engraved with a cryptic inscription, adding a layer of interest and sophistication to the task.

THE ALLURING GOLD: THE FIRST SUITOR'S TEST

The Prince of Morocco steps forward to face this intriguing test. He is confronted with the inscriptions on the caskets, each one at least as cryptic as the others. Drawn by the promise of desire inscribed on the gold casket, 'Who chooseth me shall gain what many men desire', he chooses it, hoping to find 'an angel in a golden bed'. His dreams are shattered when he finds a skull and a cryptic scroll, instead of the image of Portia. The message on the scroll serves as a harsh reminder, 'All that glisters is not gold'. With a heavy heart, he retreats, leaving Portia with a sigh of relief, uttering, 'A gentle riddance'.

SILVER'S DECEPTION: THE SECOND SUITOR'S TURN

Emboldened by his self-worth, and unaware of which casket his predecessor had chosen, the Prince of Arragon interprets the inscription on the silver casket, 'Who chooseth me shall get as much as he deserves', as a validation of his worthiness. He selects this casket.

ADDING COMPLEXITY: THE PUZZLE TAKES A TWIST

Now let's indulge in a thought experiment by introducing an intriguing layer to this complex puzzle. Suppose that, after Arragon's selection of the silver casket, Portia must open one of the remaining caskets without revealing the portrait's location. She must, therefore, open the gold casket, which she knows does not contain her likeness.

This presents Arragon with the opportunity to hold to his initial choice, the silver casket, or switch to the remaining, unopened casket made of lead.

WEIGHING THE ODDS: ARRAGON'S PROBABILITY PARADOX

If Arragon believes that Portia's knowledge of the caskets is equal to his, should he stick with his initial choice or take a chance on the unopened lead casket? His decision is far from straightforward, hinging on his interpretation of the cryptic inscriptions, his understanding of the shifting probabilities, and his perception of Portia's actions.

THE PROBABILITY PUZZLE: DECIPHERING THE GAME OF CHANCE

To understand the implications of the new development, we must first delve into the realm of probability. At the outset, Arragon's initial choice, the silver casket, had a one-third chance of being correct, assuming he has no other information. There is, therefore, a two-thirds probability that the portrait lay in one of the other two caskets.

Portia's revelation that the gold casket doesn't contain the portrait effectively shifts these odds if we can assume that she knows which of the caskets contains her portrait, and must not reveal it. The two-thirds chance, which was initially split between the gold and lead caskets, now converges entirely on the lead casket. Consequently, if Arragon changes his choice from the silver casket to the lead one, his probability of finding Portia's portrait doubles from one-third to two-thirds, other things being equal.

FATEFUL DECISION: TO SWITCH OR NOT TO SWITCH

If he dismisses the inscriptions as mere distractions and recognises the probability shift in favour of the lead casket, then switching seems like the most rational move. However, if he believes that he has deciphered the true meaning of the inscriptions, he might decide to stick with his original choice.

ARAGON'S DECISION: TO OPEN THE SILVER CASKET

Arragon is either unaware of the true probabilities or else is swayed by the cryptic clues. He chooses the silver casket. However, it only harbours disappointment. Instead of Portia's portrait, he discovers an image of a fool and a note mocking his decision, 'With one fool's head I came to woo, But I go away with two'. His self-confidence leads to his downfall, leaving him more foolish than when he first arrived.

THE POWER OF THE INSCRIPTIONS: GUIDE OR DISTRACTION?

The inscriptions on the caskets add an extra layer of uncertainty and complexity to Arragon's decision-making process. They could be seen as guides leading the suitors to the correct choice, or they could be deceptive distractions meant to confuse and mislead. The inscription on the lead casket, 'Who chooseth me must give and hazard all he hath', could be perceived as a warning of the risks involved or as a subtle hint about the potential rewards of choosing what appears to be the least valuable casket.

CONCLUSION: THE POWER OF INFORMATION

In this thought experiment, the key element is the new information introduced by Portia when she opens the gold casket. After all, she knows where the portrait is. This single action has the potential to increase significantly Arragon's chance of success. If Arragon understands and acts upon this new information, he can potentially improve his chances of selecting the correct casket from one in three to two in three. However, this seemingly simple shift in probability is complicated by the presence of other potentially influential factors, such as the cryptic inscriptions on the caskets. This makes the problem different from the basic Monty Hall decision.

He might also believe that Portia has no idea which casket contains the portrait. In that case, by opening the gold casket, she would be adding no information to what Arragon already has. He may as well be guided by any additional information he thinks he might pick up from the cryptic inscriptions. Either way, he faces a lonely but life-altering decision.

When Should We Expect Mercy?
Exploring the Death Row Problem

THE SETTING

The setting is a prison where three inmates—Amos, Bertie, and Casper—are awaiting the hangman's noose. The warden, in an act of clemency to celebrate the King's birthday, will grant clemency to just one of the three prisoners. The choice of which of these inmates to pardon is made randomly, with each name placed in a hat and drawn out. The warden now knows whom he will pardon, but the men on death row do not.

THE REQUEST

Amos makes a request to the warden. He asks the warden to name a prisoner who will NOT be pardoned, without revealing his (Amos's) own fate. If the warden has chosen Bertie to be granted clemency, he should name Casper as one of the doomed. If it's Casper who has been pardoned, the warden should name Bertie to be executed. If Amos himself is to be pardoned, the warden should simply toss a coin and name either Bertie or Casper as one of the doomed.

Amos's Request: It's essential to note here that Amos's request is based on the assumption that the warden will not reveal if Amos himself is the pardoned prisoner.

THE WARDEN'S RESPONSE

The warden agrees to the request from Amos and reveals that Casper is not the pardoned prisoner.

WHAT DOES THIS MEAN FOR AMOS AND BERTIE?

With this new information to hand, each of the prisoners can re-evaluate their chances. Initially, Amos believes that his chance of a pardon is 1/3, but with Casper out of the running, he believes that his odds of clemency have risen to 1/2. But is he right in this belief?

RE-EVALUATING THE ODDS

Initially, the odds are 1/3 for each prisoner because only one of the three prisoners is chosen at random to be pardoned. However, when the warden reveals that Casper will not be pardoned, Amos gains new information but not about his own fate. There's no new information regarding his own fate, so his chances remain as they were, at 1/3. Meanwhile, Bertie's odds of being pardoned have now increased to 2/3.

WHY ARE THE ODDS DIFFERENT?

This difference in the odds between Amos and Bertie might seem to be counterintuitive. How can they both receive the same information, yet have different survival odds? The answer lies in the warden's selection process. The warden would not have revealed Amos as the condemned prisoner due to Amos's unique request, but he might have revealed Bertie as such, instead of Casper. The fact that he doesn't name Bertie when he might have done so indicates that Bertie's chances of being pardoned have increased, while nothing has changed for Amos. Amos's belief that his odds have increased to 1/2 is a misconception.

A LARGER SCENARIO

If this still seems puzzling, consider a larger group of 26 prisoners. If Amos asks the warden to name 24 condemned prisoners in random order without revealing his own fate on any occasion, each prisoner initially has a 1/26 chance of being pardoned. But every time a doomed prisoner is named, the chance that each of the remaining prisoners (except for Amos) will be pardoned increases.

Once every prisoner but Bertie has been named as condemned, Amos's chances of survival remain at 1/26. However, Bertie's odds of being pardoned have now increased to 25/26, even though only two prisoners remain unnamed by the warden.

So, even though it might seem like Amos has a very good chance of being pardoned, the reality is that his odds have not changed and remain at 25/1, representing a chance that Amos will escape the noose of 1/26.

CONCLUSION: THE KEY TAKEAWAY

The Three Prisoners Problem highlights the importance of understanding the method by which we obtain information and its impact on the probabilities. It's a fascinating exploration of conditional probability that shows how the same piece of information can affect the chances of two individuals differently, based on the process by which that information was revealed. As such, it is a classic example of how counterintuitive probability can be, especially in situations where information is revealed in a conditional manner.

When Should We Expect to Find Gold? Exploring the Box Enigma

THE GAME AND THE PUZZLE

The Bertrand's Box Paradox, first posed by mathematician Joseph Bertrand, offers a fascinating challenge to our intuitive grasp of probability.

In Bertrand's scenario, there are three indistinguishable boxes. Each is closed. The first box contains two gold coins, while the second box holds two silver coins. The third box contains one gold and one silver coin. This setup paves the way for an exploration of probability and decision-making that might seem to challenge common sense.

THE CHOICE AND THE IMPLICATION

Imagine yourself in this setting. You randomly select one of these boxes and, without looking, you take one of the two coins from that box. As you open your hand, you see a shiny gold coin resting in your palm.

Now, the presence of the gold coin means that you didn't select the box containing the two silver coins. Thus, the box in front of you must either be the one containing two gold coins or else it is the one containing one gold and one silver coin. With this information, what is the probability that the other coin in the box is also gold?

THE INTUITIVE ANSWER AND THE SURPRISE

At first glance, the problem appears simple. Having excluded the box containing the two silver coins, we are left with two possible boxes: a box with two gold coins, and a box with one gold and one silver coin. Based on this information, we might presume that the likelihood of each box being the one we randomly selected should be equal. This presumption would lead us to the intuitive conclusion that the chance the other coin is gold stands at 1/2. Likewise, the chance that it is silver would also be 1/2. But is this intuition correct?

In fact, the truth diverges from this intuitive explanation. The correct answer to the probability that the other coin is gold is not 1/2, but 2/3. This outcome might seem to defy common sense. How could merely examining one coin influence the composition of the remaining concealed coin?

THE REVELATION AND THE TRUE ANSWER

To solve this puzzle, we need to look deeper into the details. To do so, let's imagine that each coin in the boxes has a unique label. In the gold coin box, we have Gold Coin 1 and Gold Coin 2. In the mixed box, there's Gold Coin 3 and Silver Coin 3, while the silver box holds Silver Coin 1 and Silver Coin 2.

When we initially drew a gold coin from our chosen box, three equally likely events could have occurred. We could have drawn Gold Coin 1, Gold Coin 2, or Gold Coin 3. We remain unaware of which specific gold coin we hold, but the outcomes for the remaining coin in the box vary based on this choice. If we had picked Gold Coin 1 or Gold Coin 2, the remaining coin in the box would also be gold. So, there are two chances it would be gold. However, if it was Gold Coin 3, the other coin in the box would be silver. This is one chance compared to the two chances it is gold.

When we consider these equally likely scenarios, the probability that the other coin is gold stands at 2/3, whereas the probability that it's silver is 1/3. A seemingly simple choice of coin selection reveals in this way a solution that seems to challenge our intuitive understanding of probability.

THE IMPACT OF NEW INFORMATION

Before we drew the gold coin, the probability that we had chosen the box with two gold coins was 1/3. But when we uncovered the gold coin, we didn't merely exclude the box with two silver coins, we also gathered new information. Specifically, we could have drawn a silver coin if our selected box was the one with mixed coins, yet we drew a gold coin. This fresh piece of information now means that it is twice as likely that we chose the box with two gold coins rather than the mixed one, because there were two ways this could have happened, compared to just one way if we had selected the mixed box.

A BRIEF SUMMARY OF THE BOX PARADOX

Imagine there are three boxes:

1. Box 1 contains two gold coins.
2. Box 2 contains one gold coin and one silver coin.
3. Box 3 contains two silver coins.

Initially, without any further information, the probability of choosing any one of the boxes is 1/3.

When you draw a coin and see that it is gold, you're effectively eliminating Box 3 (the box with two silver coins) from consideration because it cannot possibly be the box you chose. This leaves you with Box 1 and Box 2 as possibilities.

However, the key insight is in how we update our probabilities based on the new information that the drawn coin is gold:

- For Box 1 (with two gold coins), there are two chances of drawing a gold coin (since both coins are gold).
- For Box 2 (with one gold and one silver coin), there is only one chance of drawing a gold coin.

Therefore, given that you have drawn a gold coin, it is twice as likely that you have chosen Box 1 compared to Box 2. This updates the probabilities to:

- 2/3 chance that the box chosen was Box 1 (with two gold coins).
- 1/3 chance that the box chosen was Box 2 (with one gold and one silver coin).

CONCLUSION: THE KEY LESSON

The paradox of Bertrand's Box serves to remind us of the nuanced nature of probability, and illustrates the central importance of incorporating new information into probability calculations. Ultimately, though, it highlights the deceptive character of common intuition, encouraging us to challenge our feelings with the power of reason.

When Should We Push the Envelope?
Exploring the Exchange Paradox

A DIVE INTO THE TWO ENVELOPES PARADOX: UNRAVELLING THE ENIGMA

The 'Two Envelopes Paradox', also known as the 'Exchange Paradox', is a classic conundrum of choice and value. This deceptively simple dilemma presents us with two envelopes, each containing a certain amount of money. The rules of the dilemma are simple. One envelope contains exactly twice as much as the other. We choose an envelope, look inside, and then face the option of sticking with our original choice or switching to the other envelope.

THE ALLURE OF THE SWITCH

At first glance, the decision seems straightforward. We stand to double our money by switching. Let's say the chosen envelope contains £100. In that case, the other envelope either contains £50 (half) or £200 (double). It's tempting to switch, as we could either gain £100 or lose £50. It appears that on average we are better off switching, no matter what amount lies in the initial envelope, since we are equally likely to gain £100 as to lose £50.

The allure of the switch persists even when we do not know the envelope's contents. We could argue that if the chosen envelope holds X pounds, then the amount in the other envelope would be either 2X or 1/2X, with equal likelihood. Mathematically, this can be shown to equate to an expected value of ½ (2½X), or 5/4X, which is greater than X. On this basis, it seems a good idea to switch envelopes.

AN INFINITE DILEMMA: THE ABSURDITY OF THE SWITCH

Following this line of logic might lead us to a bewildering conclusion. Why not switch back and forth between the envelopes endlessly? If each switch supposedly increases the expected value, would we not become ever wealthier by just continually swapping envelopes?

This conclusion defies our sense of reality. We know that there's something fundamentally wrong with the idea of a perpetual money-making machine created just by swapping envelopes. Yet, where does our logic fail us?

STEPPING BACK: VIEWING THE TOTAL PICTURE

A different approach to the problem is to consider the total sum of money present in both envelopes. Let's represent this total as A. Since one envelope contains Y pounds and the other has twice as much, 2Y, we know that A equals 3Y.

If we initially picked the envelope with Y, switching to the 2Y envelope would give us an additional Y. However, if our first choice was the 2Y envelope, switching to the Y envelope would result in the loss of Y. So, there is an equal chance of gaining Y as of losing Y by making the switch. Balancing out these probabilities, we can conclude that the expected gain from switching is precisely zero.

RESOLVING THE PARADOX: FRAMING THE PROBLEM CORRECTLY

A key reason why the paradox seems so puzzling lies in how we frame the situation. The argument for switching implies that there are three possible amounts of money in play: X, 2X, or 1/2X. However, we know that there are only two envelopes, hence only two possible amounts.

By accurately framing the problem with just two amounts of money, we realise there is no expected gain or loss from switching envelopes. This remains true whether we frame the amounts as X and 2X or as X and 1/2X. Regardless, the average gain from switching equals zero.

CONCLUSION: EMBRACING THE MYSTERY OF PROBABILITY

The Two Envelopes Paradox demonstrates the often counterintuitive nature of probability and expected value. Despite the tempting initial logic suggesting a continual switch might be profitable, careful consideration reveals that without additional information there is, in fact, no inherent benefit to switching—a twist that showcases the often-mystifying appeal of mathematical reasoning.

When Should We Expect a Boy?
Exploring the Two Child Problem

THE BOY OR GIRL PARADOX

The Boy or Girl Paradox, also known as the Two Child Paradox, is a fascinating probability puzzle that challenges our intuitive understanding of probabilities. The paradox revolves around a simple scenario: a family with two children, where one of the children is known to be a boy. The question that arises is: What is the probability that the other child is also a boy? Intuitively, one might assume that the probability is 50%, as there appear to be only two possibilities: a boy or a girl, and we assume that in general a child is equally likely to be a boy or a girl. However, a more detailed analysis reveals that the correct probability is 1/3. To fully grasp the paradox and its implications, let's dive deeper into the concepts of probability and conditional probability, as well as explore various scenarios and explanations.

ANALYSING THE GENDER COMBINATIONS

To begin our analysis, let's consider all the possible combinations of genders for the two children. We can denote a boy as B and a girl as G. With these symbols, the four potential combinations of genders are:

1. **Boy–Boy (BB)**
2. **Boy–Girl (BG)**
3. **Girl–Boy (GB)**
4. **Girl–Girl (GG)**

It's important to note that each combination is equally likely, assuming an equal chance of a child being a boy or a girl.

THE PARADOX REVEALED: EVALUATING THE PROBABILITIES

Now, let's examine each combination and its implications for the Boy or Girl Paradox:

1. **Boy–Boy (BB):** This combination represents the scenario where both children are boys. Out of the four possible combinations, BB has a probability of 1/4. It can be achieved in only one way: both children being boys (BB).
2. **Boy–Girl (BG):** This combination represents the scenario where the first child is a boy and the second child is a girl. This could be based, for example, on the order in which they were born. Like BB, the BG combination also has a probability of 1/4.
3. **Girl–Boy (GB):** Similar to the BG combination, this combination also has a probability of 1/4.
4. **Girl–Girl (GG):** This combination represents the scenario where both children are girls. Out of the four possible combinations, GG has a probability of 1/4. It can be achieved in only one way: both children being girls (GG).

CONDITIONAL PROBABILITY AND THE RESOLUTION OF THE PARADOX

So, one of the two children is known to be a boy. Out of the three remaining possibilities (BB, BG, and GB), only one combination (BB) has both children being boys. Therefore, the probability of the other child being a boy is 1/3. This means that in scenarios where we know one child is a boy, the probability of the other child being a boy is 1/3, not the intuitive 1/2. The paradox arises from the fact that we often overlook the distinction between the BG and GB scenarios, treating them as a single outcome. In fact, they represent two distinct possibilities. The Boy or Girl Paradox serves as a reminder of the importance when solving probability problems of carefully analysing the given information, considering all possible outcomes, and questioning our assumptions.

EXPLORING DIFFERENT SCENARIOS AND EXPLANATIONS

To gain a deeper understanding, let's explore the Boy or Girl Paradox from different perspectives and scenarios. This will help solidify our understanding of conditional probability and shed light on why the intuitive answer of 1/2 is incorrect.

SCENARIO 1: IDENTIFYING THE BOY

Imagine you meet a man at a conference who mentions his two children and reveals that one of them is a boy. What is the likelihood that his other child is a girl? Most people would intuitively assume the probability is 1/2, but it is actually 2/3. The key to understanding this lies in the fact that we do not have information about which child, the older or the younger, is the boy. If the man had specified that the older child is a boy, then the probability would indeed be 1/2. However, since we don't have that specific information, the probability changes.

To illustrate this, let's consider the possible combinations of genders when we know one child is a boy:

1. **Older Child: Boy - Younger Child: Boy (BB)**
2. **Older Child: Boy - Younger Child: Girl (BG)**
3. **Older Child: Girl - Younger Child: Boy (GB)**

In this scenario, options 2 and 3 are equally likely. Therefore, there is a 2/3 probability that the other child is a girl, as only one out of the three possibilities (BB) has both children being boys.

SCENARIO 2: DIFFERENTIATING BETWEEN CHILDREN ALTERS PROBABILITY OUTCOMES

Any method allowing us to differentiate between one boy and another, or one girl and another, changes the probabilities. For example, if we are told that the older child is a boy, we can eliminate option 3, leaving just options 1 and 2. In this case, the probability is 1/2 that the other child is a girl, not 2/3.

Using the same logic, suppose a different scenario in which you meet a man in the park with his son and find out that he has two children, but nothing else. Well, in this case, there are only two possibilities:

1. Boy in the park—Girl at home
2. Boy in the park—Boy at home

Clearly, the probability that the other child (the child at home) is a girl now becomes 1/2.

In this case, it is location (the boy is in the park, the other child is not) rather than order of their birth that is the distinguishing characteristic.

APPLYING THE SAME CONCEPT TO A COIN TOSS

This scenario can be equated to having two coins and knowing that at least one of them is heads up. So, what's the probability of the other coin also being heads? With two coins, four outcomes are possible: Heads—Heads, Heads—Tails, Tails—Heads, Tails—Tails. After learning that at least one of the coins is Heads, we can discount the Tails—Tails possibility. We're left with three equally likely scenarios: two of these contain a Tails in the binary pair and one contains a Heads. Consequently, the likelihood that the other coin is Tails is 2/3. If, on the other hand, we are told that the first of two coins has landed heads up, what is now the chance that the second coin will land tails up? Now, it's 1/2. By introducing a distinguishing feature, such as the first child that was born or the first coin that was tossed, we change the conditional probability.

GIRL NAMED FLORIDA SCENARIO

Suppose instead we learn that one of the girls is named Florida, which is a good discriminating characteristic. How does this additional information affect the probability of the other child being a boy? Let's explore this scenario.

If you identify one of the children, say a girl named Florida, only two of the following four options exist:

1. Boy, Boy
2. Girl named Florida, Girl
3. Girl named Florida, Boy
4. Girl not named Florida, Boy

In this case, the name serves as the discriminating characteristic instead of order of birth, say, or location. Options 1 and 4 can be discarded in this scenario, leaving Options 2 and 3. In this case, the chance that the other child is a girl (almost certainly not named Florida) is 1 in 2. Similarly, the chance that the other child is a boy is also 1 in 2.

This example demonstrates how additional specific information, notably identification of a discriminating characteristic of some kind, can impact the probabilities.

VARIATIONS OF THE PARADOX

The Boy or Girl Paradox is sensitive, therefore, to the context of the problem, which can impact the solution. Subtle changes in this can lead to different solutions. It is for this reason crucial to understand the precise context and conditions when evaluating probability problems.

For example, consider two variations of the initial problem:

1. **Variation 1:** 'Mr. Smith has two children, and one of them is a boy—that's all you know. What is the probability that the other is also a boy?' In this case, the correct answer would be 1/3.
2. **Variation 2:** 'Mr. Smith has two children, and you see one of them, who is a boy. What is the probability that the other is also a boy?' In this case, the correct answer would be 1/2. By physically observing a boy, we gain additional information that distinguishes between the Boy–Girl and Girl–Boy combinations, leading to different probabilities. In this case, location is the distinguishing characteristic.

These variations highlight the importance of understanding the precise context of the problem to arrive at the correct solution.

CONCLUSION: REAL-LIFE APPLICATIONS AND IMPORTANCE

While the Boy or Girl Paradox is a theoretical puzzle, it offers valuable insights into real-world situations involving probabilities. In particular, the paradox serves as a reminder that we must be cautious when interpreting probabilities in real-life situations. It emphasises the importance of carefully considering the context, available information, and potential biases that could influence our judgment. By developing a strong foundation in probability theory, critical thinking skills, and understanding conditional probabilities, we can make more informed decisions, minimise risks, and optimise outcomes in both personal and professional contexts.

When Should We Expect a Party?
Exploring the Birthday Paradox

SIZE MATTERS

What is the minimum number of individuals that need to be present in the room for it to be more likely than not that at least two of them share a birthday? This is what the 'Birthday Problem' seeks to solve.

For the sake of simplicity, let's assume that all calendar dates have an equal chance of being someone's birthday and let's disregard the Leap Year occurrence of 29 February.

A BASIC INTUITION: ANALYSING THE ODDS

At first glance, you might think that the odds of two people sharing a birthday are incredibly low. In a group of just two people, the likelihood of them sharing a birthday is a mere 1/365. Why is that? We have 365 days in a year, hence there's only one chance in 365 that the second person would have been born on the same specific day as the first person.

Now, let's take a group of 366 people. In this case, it's certain that at least one person shares a birthday with someone else, due to the simple fact that we only have 365 possible birthdays (ignoring Leap Years).

The initial intuition may suggest that the tipping point—the group size at which there's a 50% chance of two individuals sharing a birthday—is around the midpoint of these two extremes. You may think it lies around a group size of about 180. However, the reality is surprisingly different, and the actual answer is much smaller.

THE CALCULATIONS: UNRAVELLING
THE BIRTHDAY PARADOX

To understand the concept better, we need to dig deeper into the probabilities involved. Let's consider a duo: Julia and Julian. Let's assume that Julia's birthday falls on 1 May. The chance that Julian shares the same birthday, assuming an equal distribution of birthdays across the year, is 1/365.

What about the probability that Julian doesn't share a birthday with Julia? It's simply 1 minus 1/365, or 364/365. This number illustrates the chance that the second person in a random duo has a different birthday than the first person.

Adding a third person into the mix changes things slightly. The chance that all three birthdays are different is the chance that the first two are different (364/365) multiplied by the probability that the third birthday is unique (363/365). So, the probability of three different birthdays equals (364/365)×(363/365).

As we expand the group, the calculations continue in a similar manner. The more people in the room, the greater the chance of finding at least two people sharing a birthday.

Consider a group of four people. The probability that four people have different birthdays is (364×363×362)/(365×365×365). To find the probability that at least two of the four share a birthday, we subtract this number from 1. Thus, the odds of having at least two people with the same birthday in a group of four are about 1.6%.

As the number of people in the room increases, the probability of at least two sharing a birthday grows:

- With 5 people, the probability is 2.7%.
- With 10 people, the probability is 11.7%.
- With 16 people, the probability is 28.1%.
- With 23 people, the probability is 50.5%.
- With 32 people, the probability is 75.4%.
- With 40 people, the probability is 89.2%.

THE PARADOX UNVEILED: IT'S NOT JUST ABOUT BIRTHDAYS

You might be wondering why we need just 23 people to reach a 50% chance of finding shared birthdays. This can be explained by how many possible pairs can be made in a group. In a group of 23, there are 253 unique pairs. Each of these pairs has a 1/365 chance of sharing a birthday, and all these possibilities add up. This is what makes the birthday problem so counterintuitive. Basically, when a large group is analysed, there are so many potential pairings that it becomes statistically likely for coincidental matches to occur.

This is a perfect demonstration of the concept of multiple comparisons and an example of the so-called 'Multiple Comparisons Fallacy'.

The same reasoning applies to balls being randomly dropped into open boxes. Assume there is an equal chance that a ball will drop into any of the individual boxes, and there are 365 such boxes, into which 23 balls are randomly dropped. There is an an equal chance, we ssume, that a ball will drop into any specific box. Now, there is just over a 50% chance in this scenario that there will be at least two balls in at least one of the boxes. Randomness produces more aggregation than intuition leads us to expect.

YOUR PERSONAL BIRTHDAY CHANCES: WHERE DO YOU STAND?

The reason for the paradox is that the question is not asking about the chance that someone shares your particular birthday or any particular birthday. It is asking whether any two people share any birthday.

While the birthday problem shows the increased likelihood of shared birthdays in a group, the chance that someone shares *your* birthday specifically is a different question.

In a group of 23 people, including yourself, the probability that at least one person shares your birthday is much lower than 50%—it's about 6%. This is because there are only 22 potential pairings that include you.

Even in a group of 366 people, the probability that someone shares your specific birthday is only around 63%.

CONCLUSION: THE MAGIC OF PROBABILITY AND THE BIRTHDAY PARADOX

The Birthday Paradox reveals an intriguing counterintuitive fact about probability: a group of just 23 people has a greater than 50% chance of including at least two people who share the same birthday. It sheds light on the intricacies of probability by demonstrating how many opportunities there are for matches to occur, even in seemingly small groups. For example, if you can find out the birthdays of the 22 players at the start of a football game, and the referee, more than half of the time two of them will share a birthday.

This fascinating concept has applications way beyond birthdays. It's also very important for the safety and performance of computer systems and online security. This idea helps specialists prevent and deal with issues that occur when data unexpectedly overlaps. Understanding the paradox is crucial, therefore, for those who design and secure computer systems, helping them to make these systems more reliable and efficient.

Nevertheless, it's in the social setting of parties where the paradox becomes a delightful surprise. Next time you're among friends or at any casual meet-up, consider introducing this paradox; you might just bring to life the unexpected magic of probability!

When Should We Expect to Wait? Exploring the Inspection Paradox

THE BUS STOP SCENARIO

Take the case of a bus that arrives, on average, every 20 minutes. It's not a perfect rule—sometimes the bus arrives early and sometimes it's late. But, when you calculate all the arrival times, it averages out to three times an hour, or every 20 minutes.

Now, picture yourself emerging from a side street to the bus stop, with no idea when the bus last arrived. The question that naturally arises is: how long should you expect to wait for the next bus?

Your initial thought might be, 'Well, if it's 20 minutes on average, then I should expect to wait around 10 minutes'. This would be halfway between the average intervals and would indeed be the case if the bus arrivals were perfectly spaced out. However, if you find yourself waiting longer than this, you might start to feel like the world is against you. The question then arises: are you just unlucky, or is something else at play?

This is where we introduce the concept of the Inspection Paradox.

UNRAVELLING THE INSPECTION PARADOX

The Inspection Paradox is a statistical phenomenon that reveals how our expected wait times can differ from the average times we calculate, due to the randomness of our inspections or experiences.

To illustrate this, let's look deeper into the bus scenario. The bus schedule is not as straightforward as it might seem. Remember, the bus arrives every 20 minutes on average, but not at precise 20-minute intervals. Variability changes things.

UNPREDICTABILITY IN THE BUS SCHEDULE

Consider a situation where half of the time the bus arrives at an interval of 10 minutes, and the other half at an interval of 30 minutes. The overall average remains at 20 minutes, but your experience at the bus stop will differ. If you show up at the bus stop at a random time, it's statistically more probable that you will turn up during the longer 30-minute interval than the shorter 10-minute interval.

This variation has significant implications for your expected wait time. If you land in the 30-minute interval, you can expect to wait around 15 minutes, half of that interval. If you find yourself in the 10-minute interval, you'll only wait around 5 minutes on average. However, you're three times more likely to hit the 30-minute gap, which means your expected wait time skews closer to 15 minutes than 5 minutes. On average, your expected wait time becomes 12.5 minutes, contrary to the intuitive answer of 10 minutes. This is calculated as follows: $(3 \times 15 + 1 \times 5)/4 = 50/4 = 12.5$ minutes.

IMPLICATIONS OF THE INSPECTION PARADOX

This surprising realisation is the crux of the Inspection Paradox. It essentially states that when you randomly 'inspect' or experience an event without knowing its schedule or distribution beforehand, it often seems to take longer than the average time. This isn't due to some cosmic force giving you a hard time; it's simply how probability and statistics operate in the randomness of real life.

Understanding the Inspection Paradox can fundamentally change how you interpret your everyday experiences. It's not about bad luck but rather about understanding that your perception of averages can be skewed by variability around the average.

EVERYDAY INSTANCES OF THE INSPECTION PARADOX

Once you're aware of the Inspection Paradox, you might start noticing it in various aspects of your everyday life.

EDUCATION INSTITUTION: AVERAGE CLASS SIZE

Consider an educational institution that reports an average class size of 30 students. Now, if you were to randomly ask students from this institution about their class size, you might find that your calculated average is higher than the reported 30.

Why does this happen?

The Inspection Paradox is at play here. If the institution has a range of small and large classes, you're more likely to encounter students from larger classes in your random sample. This leads to a bigger average class size in your interview sample compared to the actual average class size.

Say, for example, that the institution has class sizes of either 10 or 50, and there are equal numbers of each. In this case, the overall average class size is 30. But in selecting a random student, it is five times more likely that they will come from a class of 50 students than from a class of 10 students. So for every one student who replies '10' to your enquiry about their class size, there will be five who answer '50'. So the average class size thrown up by your survey is $5 \times 50 + 1 \times 10$, divided by 6. This equals $260/6 = 43.3$. The act of inspecting the class sizes thus increases the average obtained compared to the uninspected average. The only circumstance in which the inspected and uninspected averages coincide is when every class size is equal.

LIBRARY STUDY TIMES

Consider another scenario where you visit a library and conduct a survey asking the attendees how long they usually study. You might notice that the reported study times are generally higher than you might have expected. This can happen not because of any over-reporting but because the sample of students you survey is skewed towards those who spend longer times studying in the library. The reason is that the longer a student stays in the library, the higher the chance you'll find them there during your random survey. Short-term visitors are less likely to be part of your sample, skewing the average study time upwards.

THE RESTAURANT AND THE SUPERMARKET

You might think about the implications for other scenarios, such as restaurant wait times or queue lengths at supermarkets. For the reasons we have learned about, we might expect our individual experience of waiting to be that little bit longer than a calculation of the unobserved average.

THE PARADOX IN OTHER
REAL-LIFE SCENARIOS

Potato Digging

Why do you often accidentally cut through the biggest potato when digging in your garden? It's because larger potatoes take up more space in the ground, increasing the likelihood of your shovel hitting them.

Downloading Files

Consider the frustration when your internet connection breaks during the download of the largest file. It's because larger files take longer to download, increasing the window of time for potential connection issues to arise.

CONCLUSION: A NEW LENS

Understanding the Inspection Paradox equips you with a new lens through which to look at the world. It helps explain why your experiences might often differ from average expectations. It's simply the laws of probability and statistics unfolding in a world full of randomness. With this knowledge, you can navigate the world with more informed expectations and a greater appreciation for statistical realities.

When Should We Expect the Data to Deceive? Exploring Simpson's Paradox

UNDERSTANDING SIMPSON'S PARADOX

Understanding complex statistical phenomena can be a daunting task, especially when they seem to defy common sense. One such concept is Simpson's Paradox, a surprising phenomenon that occurs when a trend observed within several different groups of data disappears or reverses when these groups are combined. Think of it like a recipe. Individual ingredients might have distinct flavours, but when mixed together, the overall taste can be quite different. Similarly, separate sets of data may tell one story, but when combined, they can tell a completely different one.

Let's look at some examples.

SIMPSON'S PARADOX IN MEDICINE TRIALS

Suppose you're testing the effectiveness of two different types of medicine: a new drug and an old drug. Your goal is to determine which one is more effective at treating a certain condition. You administer these drugs to different groups of patients and then analyse how well each drug performs.

Let's look at a two-day medical trial comparing two drugs.

On Day 1, the new drug showed a 70% success rate in a large group, while the old drug showed an 80% success rate in a much smaller group. This makes it seem like the old drug is better.

On Day 2, the new drug, applied to a small group, was less effective than on Day 1, while the old drug, applied to a larger group, was also less effective than on Day 1. Even so, once again the old drug seems to perform better than the new drug.

However, when we combine both days' data, the new drug comes out ahead. This shift is a classic example of Simpson's Paradox.

Day 1: Initial Observations

On the first day, you test the new drug on 90 patients, and it works for 63 of them, giving a success rate of 70%. In contrast, you administer the old drug to a smaller group of ten patients, and it works for eight of them, resulting in an 80% success rate. At this point, it seems like the old drug outperforms the new one. But let's continue.

New drug: 90 patients; 63 successes. Success rate = 70%.
Old drug: 10 patients, 8 successes. Success rate = 80%.
Conclusion: Old Drug Outperforms New Drug

Day 2: More Data, More Surprises

The following day, the new drug is given to a different group of ten patients. This time, it only works for four of them, resulting in a decreased success rate of 40%. The old drug, on the other hand, is given to a larger group of 90 patients, and it works for 45 of them, indicating a 50% success rate. Once again, the old drug seems to outshine the new one.

New drug: 10 patients; 4 successes. Success rate = 40%.
Old drug: 90 patients, 45 successes. Success rate = 50%.
Conclusion: Old drug outperforms new drug.

COMBINING THE RESULTS: SIMPSON'S PARADOX AT WORK

When you merge the results from both days, however, an interesting thing happens. The new drug, which seemed less effective on each individual day, ended up working for 67 of the total 100 patients who took it, bringing the total success rate to 67%. The old drug, conversely, worked for only 53 out of 100 of its patients, resulting in a 53% success rate overall. This is contrary to what was observed on individual days and seems paradoxical. This flip is a classic example of Simpson's Paradox.

New drug: 100 patients; 67 successes. Success rate = 67%.
Old drug: 100 patients, 53 successes. Success rate = 53%.
Conclusion: New drug outperforms old drug.

EXPLAINING THE PARADOX

The paradox in our medical trial example is heavily influenced by the size of the groups tested each day.

If we combine the results, larger group sizes on different days skew the overall success rate, revealing the paradox. The success rates are important, but the size of the groups being compared is crucial to understanding why the paradox occurs.

EXPLORING SIMPSON'S PARADOX IN DRUG EFFICACY TRIALS

Let's delve deeper into Simpson's Paradox using another example. Suppose this time you're comparing a real drug to a placebo, a sugar pill, to see if the real drug can help patients recover from a specific illness.

You arrange the patients into four distinct age groups: *elderly adults* (*Group A*), *middle-aged adults* (*Group B*), *young adults* (*Group C*), and *children* (*Group D*).

The drug's effectiveness is measured by the proportion of patients in each group who recover from their illness within two days of taking the medication.

THE SUGAR PILL EXPERIMENT

First, let's take a look at the sugar pill group.

You distribute the sugar pill to different proportions of the four age groups:

Group A has 20 elderly adults, Group B has 40 middle-aged adults, Group C has 120 young adults, and Group D has 60 children.

The sugar pill helps 10% of the elderly (Group A), 20% of the middle-aged adults (Group B), 40% of the young adults (Group C), and 30% of the children (Group D).

To calculate the overall success rate, you add up the number of successful recoveries across all the groups (2 from Group A, 8 from Group B, 48 from Group C, and 18 from Group D) and divide by the total number of patients (240). The result is 76 successful recoveries out of 240 trials, giving an overall success rate of approximately 31.7%.

Group A: 20 elderly adults; 2 successes. Success rate = 10%.
Group B: 40 middle-aged adults; 8 successes. Success rate = 20%.
Group C: 120 young adults; 48 successes. Success rate = 40%.
Group D: 60 children; 18 successes. Success rate = 30%.
Total: 240 trials; 76 successes. Success rate = 31.7%.

THE REAL DRUG EXPERIMENT

Next, let's look at the group given the real drug. This time, the group sizes are different: Group A has 120 elderly adults, Group B has 60 middle-aged adults, Group C has 20 young adults, and Group D has 40 children.

The real drug helps 15% of the elderly (Group A), 30% of the middle-aged adults (Group B), 90% of the young adults (Group C), and 45% of the children (Group D).

Again, to get the overall success rate, you add up the number of successful recoveries (18 from each group) and divide by the total number of patients (240). This time, the result is 72 successful recoveries out of 240 trials, resulting in an overall success rate of approximately 30%.

Group A: 120 elderly adults; 18 successes. Success rate = 15%.
Group B: 60 middle-aged adults; 18 successes. Success rate = 30%.
Group C: 20 young adults; 18 successes. Success rate = 90%.
Group D: 40 children; 18 successes. Success rate = 45%.
Total: 240 trials; 72 successes. Success rate = 30%.

A PARADOX EMERGES

At first glance, it seems that the sugar pill outperformed the real drug. After all, the overall success rate was higher for the sugar pill (31.7%) than for the real drug (30%). But if we examine the data more closely, we find that the real drug had a higher success rate within each age group.

So, why does the overall success rate favour the sugar pill, even though the real drug was more effective in every age category? The paradox again arises due to the different group sizes and composition.

For example, the group that took the sugar pill had a disproportionately large number of young adults (Group C). This demographic typically has higher natural recovery rates, skewing the overall success rate of the sugar pill upwards. On the contrary, the group that took the real drug had a higher proportion of elderly adults (Group A), who typically have lower recovery rates, leading to a lower overall success rate for the real drug.

A MATTER OF LIFE AND DEATH

A real-world example of Simpson's Paradox in action can be seen in the context of the COVID-19 pandemic, specifically relating to a report published in November 2021 by the Office for National Statistics (ONS). It was titled 'Deaths involving COVID-19 by vaccination status, England: deaths occurring between 2 January and 24 September 2021'.

The raw statistics showed death rates in England for people aged 10–59, listing vaccination status separately. Counterintuitively, the statistics showed that the death rates for the vaccinated in this age grouping were greater than those for the unvaccinated. These numbers were heavily promoted and highlighted on social media by anti-vaccine advocates, who used them to argue that vaccination increases the risk of death.

This claim was contrary, though, to efficacy and effectiveness studies showing that COVID-19 vaccines offered strong protection.

A CLOSER INSPECTION

Closer inspection of the ONS report reveals that over the period of the study, from January to September 2021, the age-adjusted risk of death involving COVID-19 was 32 times greater among unvaccinated people compared to fully vaccinated people. So how can we square this with the raw data? This is where Simpson's Paradox comes in.

The paradox in the ONS statistics arises specifically because death rates increase dramatically with age, so that at the very top end of this age band, for example, mortality rates are about 80 times higher than at the very bottom end. A similar pattern is observed between vaccination rates and age. For example, in the 10–59 data set, more than half of those vaccinated are over the age of 40.

Those who are in the upper ranges of the wide 10–59 age band are, therefore, both more likely to have been vaccinated and also more likely to die if infected with COVID-19 or for any other reason, and vice versa. Age is acting, in the terminology of statistics, as a confounding variable, as it is positively related to both vaccination rates and death rates. To put it another way, if you are older, you are more likely to die in a given period, and you are also more likely to be vaccinated. It is age that is driving up death rates not vaccinations. Without vaccinations, deaths would have been hugely greater from COVID-19.

STATISTICAL LITERACY

If we break down the band into narrower age ranges, such as 10–19, 20–29, 30–39, 40–49, and 50–59, the counterintuitive headline finding immediately disappears. In each age band, the death rates of the vaccinated are very much lower than those of the unvaccinated. This also applies in the higher age bands—60–69, 70–79, and 80 plus. The key point is that age is a crucial factor that must be considered when analysing the risk of death and the impact of vaccinations.

In this way, misrepresentation of statistics can have potentially devastating consequences for the lives of millions around the world. Statistical literacy is a real superpower in the global quest to protect and save these lives.

GUIDELINES AND STRATEGIES

Disaggregate the Data

Break Down Data into Subgroups: Disaggregating data by relevant subgroups (e.g. age, gender, region) can reveal underlying trends that the aggregated data might mask.

Question Initial Assumptions

Challenge Averages: Averages can be misleading. Always question what an average is concealing. Is it masking a wide distribution or skewing because of outliers?

Seek Out Hidden Variables (Confounding Variables)

Identify Potential Confounders: Simpson's Paradox often arises due to the presence of hidden variables that influence both the predictor and outcome variables.

Use Visual Data Exploration

Plot Your Data: Visualising your data can help identify patterns, trends, and anomalies. Graphs can help spot where the trend within subgroups differs from the aggregated trend, potentially signalling Simpson's Paradox.

CONCLUSION: RESOLVING SIMPSON'S PARADOX

Understanding the factors behind Simpson's Paradox allows us to make much better sense of our data. Whether in stylised examples or in the real world of a global pandemic, the paradox underscores the importance of careful data analysis, particularly when dealing with grouped data. By taking account of the sizes and characteristics of different groups, we can navigate the potential pitfalls of Simpson's Paradox and learn how to draw more informed conclusions. In a very real sense, millions of lives could depend on an understanding of this statistical reality.

When Should We Expect Everyone to Gain? Exploring the Will Rogers Phenomenon

CLARIFYING THE CONCEPT

The Will Rogers Phenomenon occurs when moving an element from one group to another increases the average of both groups. It's named after the comedian Will Rogers, who joked that people moving from Oklahoma to California raised the intelligence of both states.

Imagine moving a jigsaw piece from one box to another and somehow making both jigsaw puzzles look more complete. That's the kind of surprising outcome the Will Rogers Phenomenon describes in the world of statistics.

ILLUSTRATIVE EXAMPLE

Let's say we're looking at two groups based on a medical condition. Group A has the condition, while Group B doesn't.

Initially, Group A has a lower average life expectancy than Group B. But when one individual from Group B, who has a higher expectancy than Group A's average but lower than Group B's, is correctly diagnosed with the condition and moved to Group A, both groups' averages increase.

This might seem odd because we haven't changed anyone's life expectancy; we've just reclassified one person. Yet, the averages for both groups go up.

Consider, for example, a study analysing the life expectancies of six individuals. We assess their life expectancies one by one and find that the first two individuals have a life expectancy of 5 and 15 years, respectively. They have been diagnosed with a specific medical condition. The remaining four individuals have life expectancies of 25, 35, 45, and 55 years, but they do not have the diagnosed condition. Consequently, the average life expectancy for those with the condition is 10 years, while for those without the condition, it is 40 years.

THE IMPACT OF IMPROVED DIAGNOSIS

Now, let's suppose that advances in diagnostic medical science allow us to identify one of the individuals with a life expectancy of 25 years as actually having the medical condition, which was previously missed. This discovery prompts us to move this person from the undiagnosed group to the diagnosed group.

As a result, the average life expectancy within the group diagnosed with the condition increases from 10 to 15 years. The calculation for the new average is $(5+15+25)$ divided by 3. Simultaneously, the average life expectancy of those not diagnosed with the condition also rises by five years, from 40 to 45 years. The calculation for this new average is $(35+45+55)$ divided by 3.

THE ILLUSION OF CHANGE

Upon observing this scenario, we might be puzzled as to how moving a single data point can cause both groups' averages to increase. The Will Rogers Phenomenon provides an explanation.

In this case, the data point being moved (the individual with a life expectancy of 25 years) is below the average of the group it is leaving, which is 40 years. Yet, it is above the average of the group it is joining, which is 10 years. This creates the illusion of improvement in both groups' averages, despite there being no change in the actual values.

ADDITIONAL EXAMPLES

This phenomenon isn't limited to medical data. For example, in education, say we move a student who's not doing well in a class of high achievers to a class with lower overall scores. Suddenly, both classes seem to do better on average, even though the student's performance hasn't changed.

Suppose again there are two schools, School A and School B, with average test scores of 70% and 80%, respectively. School B now decides to send some of its lower-performing students (scores of below 80% but above 70%), while retaining its higher-performing students.

As a result, both schools' average scores increase. This occurs because the students transferred from School B to School A have scores below School B's average but above School A's average.

This example highlights how the Will Rogers Phenomenon can manifest itself in different domains, influencing various statistical analyses and interpretations.

THE ROLE OF CONTEXT

Understanding the Will Rogers Phenomenon is crucial for individuals working with statistics and data analysis. It emphasises the significance of considering context and carefully interpreting statistical results, particularly when dealing with group comparisons.

By being aware of this phenomenon, we can avoid misconstruing statistical changes as genuine improvements or deteriorations in the underlying data. It reminds us that when data points move between groups, the resulting changes in averages may not reflect true progress but rather the consequences of shifting data categorisations.

APPLICATION TO REAL-WORLD SCENARIOS

Public Health Policy

In public health, the Will Rogers Phenomenon can have profound implications, particularly in the reporting and interpretation of disease rates and the effectiveness of interventions. For instance, if a new diagnostic technique becomes available that identifies milder cases of a disease previously undetected, the overall survival rate of the diagnosed population may increase—not necessarily because the treatment has improved, but because the cohort now includes less severe cases. This can lead to the false conclusion that a new drug or treatment is more effective than it actually is, potentially influencing funding allocations, treatment protocols, and patient care strategies without genuine improvements in treatment efficacy.

Education Reforms

In the education sector, policy decisions are often influenced by the performance metrics of schools and universities. If educational standards change, causing students with lower grades to be reclassified from one performance category to another, it may appear that both the higher and lower performing groups have improved their average scores. This could lead to misleading conclusions about the success of new educational policies or teaching methods. For example, if a new grading policy causes borderline students to be classified into a lower-performing group, it might artificially inflate the average performance of both the higher and lower groups, leading to misguided policy decisions based on perceived improvements.

Economic Analysis

In economics, the phenomenon can impact the analysis of income data and the evaluation of economic policies. For example, if a government implements a new tax bracket that reclassifies some of the lower earners from the middle-income bracket to the lower-income bracket, it could appear that the average income in both brackets has increased. This could be misinterpreted as economic improvement resulting from the policy, leading to skewed analyses and subsequent policy decisions that do not accurately address the underlying economic conditions.

Environmental Policy

Consider the assessment of air quality in different regions. If new, more sensitive measuring techniques are employed that classify moderately polluted areas as highly polluted, it may appear that the average pollution levels in both the moderately and highly polluted categories have decreased. This could lead to incorrect conclusions about the effectiveness of environmental regulations and misdirected resources, impacting public health and environmental protection efforts.

Crime Statistics

Changes in the classification of crimes can lead to misunderstandings of crime trends. If, for example, certain types of thefts are reclassified, it might appear that both low-level and high-level crime rates have decreased, when in reality, only the classification criteria have changed. This can affect public perception, policy formulation, and resource allocation in law enforcement.

By providing these expanded examples, we can see how the Will Rogers Phenomenon extends far beyond statistical curiosity, affecting a wide range of important decisions in public health, education, economics, environmental policy, and criminal justice. Understanding this phenomenon is crucial for policymakers, educators, economists, and the public to avoid misinterpretations that can lead to significant real-world consequences.

CONCLUSION: SHEDDING LIGHT ON DATA PRESENTATION

Understanding the Will Rogers Phenomenon is crucial in fields like medicine, education, and any area where data is grouped and compared. It shows us that moving data around can create misleading impressions of improvement or decline.

This phenomenon highlights the need for careful data analysis and interpretation. It's a reminder that in statistics, the way we group and move data can tell different stories.

By understanding this effect, we can approach statistical analyses more critically, ensuring that we interpret changes in averages within the appropriate context.

Statistical literacy is vital for making informed decisions and avoiding misinterpretations. By embracing the complexities of statistical phenomena like the Will Rogers Phenomenon, we can become better equipped to navigate the intricacies of data-driven knowledge in our increasingly data-centric world.

A Question of Choice

3

When Should We Defy the Odds?
Exploring Newcomb's Paradox

A THOUGHT EXPERIMENT

Newcomb's Paradox, also known as Newcomb's Problem, is a thought experiment involving a choice between two boxes: one transparent containing $1,000 and one opaque that may contain nothing or $1 million. The twist? A highly accurate Predictor has already decided what's in the opaque box based on what it thinks you will choose. It is a dilemma first proposed by William Newcomb, a theoretical physicist, and popularised by philosopher Robert Nozick.

THE SETTING OF NEWCOMB'S PARADOX

In this setting, the simple choice is between taking both boxes or just taking the opaque box. The Predictor, which is well known for its accuracy, determines the content of the opaque box based on a prediction about your decision. If the Predictor forecasts you will take both boxes, it places nothing in the opaque box. On the other hand, if it predicts that you will only take the one box, a sum of $1 million will be deposited inside. By the time you make your decision, the Predictor's choice about the opaque box's content has already been made. So, should we take both boxes or just the one opaque box? You could also change the amounts to update to modern day prices or in some other way and ask yourself whether it changes anything.

DOI: 10.1201/9781003402862-3

TWO-BOXERS VS. ONE-BOXERS: THE GREAT DEBATE

Essentially, Newcomb's Paradox has divided people into two distinct camps, each adhering to a different way of looking at the problem. These factions, known as 'two-boxers' and 'one-boxers', represent different facets of decision-making theory and reflect different approaches to rational choice.

TWO-BOXERS: THE DOMINANCE PRINCIPLE AND CAUSAL DECISION THEORY

Two-boxers advocate that the most rational decision is to take both boxes. This perspective aligns with the principle of dominance in decision theory, which states that if one action produces a better outcome than another in every possible scenario, then that action should be chosen. In the case of Newcomb's Paradox, two-boxers argue that the Predictor's decision—having already determined the content of the opaque box before you choose—cannot be influenced by your subsequent choice. This means that choosing both boxes can't make you worse off. In the worst-case scenario, you have the guaranteed $1,000 from the transparent box, and in the best-case scenario, you could walk away with an additional $1 million if the Predictor failed in its prediction.

Two-boxers also fundamentally subscribe to causal decision theory. They reason that since your decision doesn't cause a change in the already-decided contents of the opaque box, it's only rational to maximise the guaranteed gains, which means taking both boxes. This standpoint portrays the logic of irreversibility, where past events (the Predictor's decision) cannot be influenced by future actions (your choice).

ONE-BOXERS: EVIDENTIAL DECISION THEORY AND TRUSTING THE PREDICTOR

Conversely, one-boxers argue for a different interpretation of rationality. They propose that the sensible decision, given the Predictor's uncanny accuracy, is to take only the opaque box. They reason that, although the contents of the box have been decided, the Predictor's ability to forecast accurately makes it likely that the opaque box contains the $1 million if you choose it alone.

One-boxers point to the track record: almost every participant who opted for two boxes found the opaque box empty, while the opposite was true for those who took only the opaque box. Hence, they argue that it's not about changing the past, but about leveraging the evidence that shows a strong correlation between the decision to pick one box and winning the million dollars.

In essence, one-boxers align with evidential decision theory, which suggests that we should make decisions based on the best available evidence for the desired outcome. In the context of Newcomb's Paradox, taking only the opaque box is based on what has happened in the past to those who took one box and two boxes, respectively.

In this way, the paradox challenges our notions of causality and rational decision-making. Can our current choice affect a decision that's already been made? Or does the Predictor's accuracy mean it's better to trust the pattern of past outcomes?

SPLIT DECISION

The paradox thus splits decision-makers into two groups: 'two-boxers' and 'one-boxers', each advocating for a different decision based on their own logic.

Two-boxers argue that the rational decision is to take both boxes. As the Predictor's decision about the content of the opaque box is already determined before you choose, your choice can't change the contents. This implies that no matter what, you won't be worse off taking both boxes. The least you can get is the $1,000 from the transparent box, and at best, you could get an additional $1 million if the Predictor predicted incorrectly.

On the other hand, one-boxers argue that the sensible decision, considering the Predictor's near-perfect track record, is to take only the opaque box. They point out that almost everyone who has taken two boxes has found the opaque one empty, while almost everyone who took only the opaque box won the million dollars. Thus, based on the evidence, it seems sensible to become a one-boxer.

The decision-making here presents a fascinating conflict between reason (which seemingly lacks supporting evidence) and evidence (which seemingly lacks rational explanation). It essentially raises the question: should we trust the evidence of a well-documented pattern or rely on the rational logic of decision-making?

CAUSALITY: THE PREDICTOR AND THE FUTURE

The first crucial point to clarify in Newcomb's Paradox is the nature of causality at play. The scenario eliminates any notion of backward causality or retro-causality; your choice does not affect the Predictor's prior decision nor alter the content of the opaque box. This stipulation aligns with our typical understanding of time's arrow: the future does not influence the past.

The Predictor's decision is a genuine prediction and doesn't involve any time-travelling. It infers your choice before you make it, but it doesn't 'react' to your decision.

PREDICTABILITY: UNRAVELLING THE ACCURACY OF THE PREDICTOR

The Predictor's high accuracy complicates the decision-making process. If you tend to be a two-boxer, you might think that it doesn't matter what the Predictor has foreseen as the decision has already been made. Conversely, if you lean towards one-boxing, you might believe that the Predictor has probably predicted this and filled the opaque box with the million dollars. The paradox then becomes less about the boxes you choose and more of a high-stakes mind game where you try to leverage the Predictor's uncanny accuracy for your gain.

IDENTITY: THE PERSON YOU CHOOSE TO BE

This leads to another fascinating dimension: the intersection of predictability and identity. If the Predictor can predict your decision based on your inherent nature, then maybe the real strategy lies in manipulating your own disposition to game the system. The question then evolves from 'which box should you choose?' to 'who should you choose to be?'

In essence, if you aspire to secure the million dollars, the optimal strategy might be to become the type of person who would always choose one box. The paradox suggests that by firmly committing to this decision, you make it likely for the Predictor to foresee this choice and fill the opaque box accordingly.

The role of the Predictor, therefore, not only tests our understanding of causality and predictability, but it also nudges us to introspect on the role our identity plays in decision-making. It prompts us to consider the potential power of a self-fulfilling prophecy, where our decision to be a certain 'type' of person may indeed lead to the desired outcome. Thus, Newcomb's Paradox elegantly encapsulates the intricate interplay between causality, predictability, and personal identity in shaping our choices and their consequences.

CONCLUSION: ONE BOX OR TWO?

The question remains: why leave the sure $1,000 in the transparent box when the content of the opaque box is already decided? Why not take both? This question is at the heart of Newcomb's Paradox. The paradox doesn't necessarily dictate a 'correct' decision.

Instead, it presents a problem that forces you to rethink rationality, predictability, and decision-making. It also highlights the complexity and paradoxical nature of decision-making when dealing with highly reliable predictors. Ultimately, though, the decision rests with you. Would you take one box or two?

When Should We Rethink Probability? Exploring the Sleeping Beauty Problem

A THOUGHT EXPERIMENT

The Sleeping Beauty Problem is a thought experiment that challenges our understanding of probability. It involves Sleeping Beauty, a coin toss, and a scenario where her memory is erased, leading to a debate between two main schools of thought: 'halfers' and 'thirders'.

THE SLEEPING BEAUTY EXPERIMENT

The experiment plays out as follows: Sleeping Beauty participates in an experiment, starting on a Sunday. The course of the experiment depends entirely on the outcome of a fair coin toss. If it lands heads, Beauty will be woken and interviewed only on Monday. If it lands tails, she will be awakened and interviewed on both Monday and Tuesday. On each occasion, she is asked what chance she assigns to the coin having landed heads. After she answers, she is put back into a sleep with a drug that erases her memory of that awakening. The experiment in any case finishes on Wednesday, with Sleeping Beauty waking up without an interview.

In other words, Sleeping Beauty participates in a coin toss experiment. If the coin lands heads, she is woken and interviewed only on Monday. If tails, she is woken on both Monday and Tuesday, with each awakening followed by memory erasure. She is asked each time about the likelihood of the coin having landed heads.

PROBABILITY PARADOX: HALFERS VS. THIRDERS

When presented with this experiment, two primary interpretations of how Sleeping Beauty should calculate the probability emerge. Halfers propose that since the coin is only tossed once and no new information is collected by Beauty, she should assert a 1 in 2 chance that the coin landed heads. On the contrary, Thirders argue that from Sleeping Beauty's standpoint, there are three equally likely scenarios, two of which involve the coin landing tails and one after a heads. Specifically, these are:

1. It landed heads, and it is Monday.
2. It landed tails, and it is Monday.
3. It landed tails, and it is Tuesday.

Therefore, Thirders suggest that whenever she wakes up, she should assign a 1 in 3 chance to the coin having landed heads.

THE BETTING FRAME: DETERMINING FAIR ODDS IN THE SLEEPING BEAUTY PROBLEM

One potential strategy for deciphering this complex issue is by considering it in terms of fair betting odds. For instance, if Sleeping Beauty were offered odds of 2 to 1 (£1 to win a net £2) that the coin landed heads, should she take the bet?

The best way to look at this is to think about what would happen if she accepted the 2 to 1 odds each time she woke up. If the coin toss results in heads, she'd be woken up once, bet £10, and profit £20. But if the coin lands on tails, she'd be woken up twice, place two £10 bets (a total of £20) and lose both times.

Her 'average' result with this betting strategy would be to break even. This implies that 2 to 1 represent the correct odds. These odds (£1 win to win a net £2) are consistent with a probability of 1/3. So, using this betting test, when Beauty wakes up, she should think there's a 1 in 3 chance that the coin landed on heads. This supports the 'Thirder' case.

SHIFTING PROBABILITIES: FROM UNCONDITIONAL TO CONDITIONAL

A critical step in unravelling this puzzle involves an examination of the 'prior probability'. This is the probability assigned before the collection of any new information. If asked to estimate the likelihood of a fair coin landing heads without any additional conditions, Beauty should answer 1/2. However, with added information, the question can be reformulated into estimating the probability of her waking as a result of the coin landing heads. Here, thirders would argue for a 1/3 probability. So, what information does Beauty actually have when she goes to sleep that Sunday, and how does that affect the prior probability that she should assign to the coin landing heads? Bear in mind, though, that the coin is only tossed once, and it is a fair coin.

CONCLUSION: HOW THE SLEEPING BEAUTY PROBLEM COMBINES CHANCE AND DEEP THOUGHT

The Sleeping Beauty Problem is more than a statistical puzzle; it's a probe into the nature of information and observation. It shows that our understanding of probability can significantly shift based on the framing of the question and the information available to us. Indeed, it shakes up how we think and makes us wonder about what 'information' really is. This serves as a powerful reminder that the real world, like the Sleeping Beauty Problem, doesn't always have easy or clear-cut answers. The more we dig into this mind-bending problem, the more we may be able to learn from it. In the next section, we peer deeper into the problem, exploring the very nature of existence and observation.

When Should We Trust Our Intuition? Exploring an Existential Coin Toss

THE EXISTENTIAL COIN TOSS

The Existential Coin Toss is a thought experiment designed to explore the nature of probability and existence. In this setting, you are asked to imagine a scenario where the existence of two distinct worlds is determined by the toss of a coin.

THE SET-UP

World A (Heads): This world is inhabited by a single individual with a black beard.

World B (Tails): This world contains two individuals, one with a black and the other with a brown beard.

Waking up in darkness in one of these worlds without prior knowledge of your world or beard colour, what probability would you assign to the coin having landed on Tails, thus placing you in World B?

The answer hinges on your basic assumptions about existence.

THE SELF-SAMPLING ASSUMPTION (SSA)

This approach encourages us to think of ourselves as a random selection from all entities that could have been us—our 'reference class'. Consequently, we are a randomly selected bearded individual, with an equal chance of living in World A (Heads) or World B (Tails). If in World B, there's a 50-50 probability of sporting either a black or brown beard.

But what happens when the light comes on and you see a black beard? Now, the probability of being in World A, where the sole inhabitant has a black beard, increases. Given the choice between World A (100% black beard chance) and World B (50% black beard chance), the likelihood of residing in World A is twice as much, making it a 2/3 chance the coin landed Heads.

THE SELF-INDICATION ASSUMPTION (SIA)

This alternative perspective suggests that you are twice as likely to be in a world where two observers exist than in a world with just one observer. Thus, you might lean towards World B, in which there are two observers, giving it twice the likelihood (2/3 chance) as World A (1/3 chance). But, once the lights are on and your beard is revealed to be black, the probability of living in World B reduces to 1/2, the same as the probability of living in World A, since your existence is confirmed in a scenario where both worlds have equal chances of your specific condition.

IMPLICATIONS

The contrasting perspectives of SSA and SIA in the Existential Coin Toss thought experiment illustrate the complexities in assigning probabilities to our own existence. The assumptions we choose significantly influence our conclusions about our likelihood of existing in one world versus another. This thought experiment not only sheds light on philosophical debates surrounding conditional probability but also challenges our understanding of existence and identity in uncertain contexts.

UNRAVELLING THE SLEEPING BEAUTY PROBLEM

The Sleeping Beauty Problem puts the Self-Sampling Assumption (SSA) and Self-Indication Assumption (SIA) to the test.

As we saw in the previous section, Sleeping Beauty volunteers for an experiment where she goes to sleep. A fair coin will be tossed to determine the next steps:

- If the coin lands on **Heads**, she will be awakened once (on Monday).
- If it lands on **Tails**, she will be awakened twice (on Monday and Tuesday).

On both awakenings, she has no memory of any previous awakenings, and thus can't tell which day it is or how many times she's been awoken. When she wakes, she's asked: 'What chance do you assign to the proposition that the coin landed Heads?'

ADOPTING THE SELF-SAMPLING ASSUMPTION (SSA)

Under this assumption, Sleeping Beauty would argue that there's a 50-50 chance the coin landed on Heads or Tails, as those are the only two possible outcomes from a fair coin toss. This perspective doesn't change upon waking. Only if she's told that it's her second awakening (which means it's Tuesday and the coin must have landed Tails) will she change her belief to 100% Tails and 0% Heads.

ADOPTING THE SELF-INDICATION ASSUMPTION (SIA)

Under this assumption, Sleeping Beauty considers the number of observer-moments—instances when she is awake and capable of making an observation. There are two such observer-moments if the coin lands Tails (one on Monday and one on Tuesday), but only one if the coin lands Heads (on Monday).

From this viewpoint, Sleeping Beauty would reason there's a 1/3 chance the coin landed Heads and a 2/3 chance it landed Tails. This is because there are three observer-moments in total (Monday on Heads, Monday on Tails, and Tuesday on Tails), and each one is equally likely. So the coin landing Tails (which creates two observer-moments) is twice as likely as the coin having landed Heads (which creates only one observer-moment).

In summary, the Sleeping Beauty Problem involves SSA and SIA to determine probabilities based on the number of awakenings. Here, the SSA leads to a 50% chance of heads, while the SIA suggests a 1/3 probability, due to more observer-moments under tails.

Thus, the SSA and SIA lead to different conclusions in the Sleeping Beauty Problem, just as in the God's Coin Toss problem. The correct approach remains a topic of debate among philosophers and statisticians, reflecting broader inquiries ino how we interpret probability and make decisions under uncertainty.

DILEMMA OF THE PRESUMPTUOUS PHILOSOPHER

The Presumptuous Philosopher Problem introduces a critical examination of the Self-Indication Assumption (SIA) by presenting a scenario where SIA seems to lead to counterintuitive or problematic conclusions.

Consider a situation where scientists are evaluating two theories, each equally supported by prior evidence. Theory 1 predicts the existence of a universe with a million times more observers than Theory 2, but evidence from a particle accelerator strongly supports Theory 2. Despite the empirical evidence supporting Theory 2, philosophers using the SIA can insist that Theory 1 is much more likely to be correct. Look at it this way. You exist, and Theory 1 makes your existence a million times more likely than Theory 2, because there are a million times more observers that exist if Theory 1 is true than if Theory 2 is true. To put it another way, given the fact that you exist, a case can be made for supporting a theory that proposes that a very large number of observers exist over a theory that proposes a much smaller number, even if empirical evidence strongly contradicts it.

But should the sheer number of potential observers really sway our belief in a theory, especially when faced with concrete evidence to the contrary? Critics argue that this perspective can lead to presumptuous conclusions, hence the name of the problem. In this way, the Presumptuous Philosopher Problem highlights the tension between this approach and traditional evidence-based reasoning,

CONCLUSION: CHOOSING OUR ASSUMPTIONS

These thought experiments illustrate the complexity of probability and existence. They challenge us to ponder which assumptions align with our intuition, and how reliable our intuition is in such abstract scenarios.

When Should We Expect the Final Curtain? Exploring the Doomsday Argument

CONTEMPLATING OUR EXISTENTIAL PREDICAMENT

The Doomsday Argument is a statistical and philosophical approach predicting humanity's potential end. It uses principles of probability to suggest that humanity might be closer to its demise than we commonly believe.

PROBABILITY AND ITS IMPLICATIONS

Imagine attempting to estimate your enemy's tank count. The tanks are sequentially manufactured, starting from one. You uncover serial numbers on five random tanks, all being under 10. In such a scenario, an intuitive grasp of probability would lead you to believe that your enemy doesn't possess a large number of tanks. However, if you stumble upon serial numbers stretching into the thousands, your estimate would justifiably swing towards a much larger count.

In another scenario, consider a box filled with numbered balls, which can either contain ten balls (numbered 1–10) or ten thousand balls (numbered 1–10,000). If a ball drawn from the box reveals a single-digit number, such as seven, it is reasonable to assume that the box is much more likely to contain ten balls than ten thousand.

INVOKING THE MEDIOCRITY PRINCIPLE AND COPERNICAN PRINCIPLE

The tank and numbered balls examples tie closely to the concept of mediocrity, as captured in the 'mediocrity principle'. This principle suggests that initial assumptions should lean towards mediocrity rather than the exceptional. In other words, we are more likely to encounter ordinary circumstances rather than extraordinary ones.

The Copernican principle dovetails with the mediocrity principle. It argues that we are not privileged or exceptional observers of the universe. This principle is rooted in Nicolaus Copernicus's 16th-century finding that Earth does not occupy a central, special position in the universe.

GOTT'S WALL PREDICTION

Astrophysicist John Richard Gott took the Copernican principle to heart during his visit to the Berlin Wall in 1969. Lacking specific knowledge about the Wall's expected lifespan, Gott took the position that his encounter with the Wall did not occur at any special time in its existence.

This assumption allowed him to estimate the future lifespan of the Wall. If, for instance, his visit was precisely halfway through its life, the Wall would stand for another eight years. If he visited one-quarter into its life, the Wall would stand for another 24 years. If visiting it three-quarters along its timeline, the future would be one-third of its past. Because half of its existence is between these two points (75% minus 25% is 50%), there was a 50% chance that it would last a further period between one-third and three times its current existence. Based on its age when he observed it in 1969 (eight years), Gott argued that there was a 50% chance that it would fall in between 8/3 years (2 years, 8 months) and 8×3 (24) years from then.

The Berlin Wall fell in 1989, 20 years after Gott's visit and roughly 28 years after it was built. This bolstered Gott's confidence in the Copernican-based method of making predictions, which he termed 'Copernican time horizons'.

The implications of Gott's Wall are far-reaching. They suggest that we could potentially apply the Copernican principle to make predictions about other systems where we have little information about their total lifespan. For example, it could be applied to predict the lifespan of a company, the duration of a war, or the longevity of a species, among many other things.

However, it's essential to acknowledge the limitations of this method. It is predicated on the assumption that there is nothing special about the moment of observation, an assumption that may not hold true in many scenarios. Despite these limitations, Gott's approach represents a fascinating application of the Copernican principle to real-world events, demonstrating how our position in time, just as in space, can be used to gain insights about the world around us.

THE LINDY EFFECT AND ITS LIMITATIONS

Gott's method finds resonance with the 'Lindy effect', the name of which is derived from a New York delicatessen, famous for its cheesecakes, which was frequented by actors playing in Broadway shows. It suggests that a show that had been running for three years could be expected on average to last for about another three years.

However, the Lindy effect has limitations. It breaks down when applied to processes like biological ageing. For instance, a human who has lived for 100 years is very unlikely indeed to live another 100 years. The factors influencing human lifespan are far from random, rendering the Lindy effect ineffective for such predictions.

FROM COPERNICAN PRINCIPLE TO DOOMSDAY ARGUMENT

The Doomsday Argument employs Gott's idea to estimate the Doomsday date for the human race. Applied to humanity, the argument contends that if we consider humanity's entire history, we should statistically find ourselves somewhere around the middle of that history in terms of the human population. If our population continues to grow exponentially, this suggests that humanity has a relatively short lifespan left, potentially within this millennium.

ESTIMATES AND PROJECTIONS

This projection takes into account the fact that there have been approximately 110 billion humans on earth to date, 7% of whom are alive today. Following demographic trends forward and estimating how long it will be for a further 110 billion humans to be born, the Doomsday Argument anticipates humanity's timeline is likely to end well within this millennium.

DEBATE AND CRITICISMS

The Doomsday Argument is not without its critics. Some argue that humanity will never go extinct, while others highlight that the argument's assumptions might not hold true, such as the assumption that humans are at the midpoint of our existence timeline. Others claim that the argument fails to account for future scientific and technological developments that might significantly extend, or perhaps foreshorten, humanity's lifespan.

CONCLUSION: THE FATE OF HUMANITY?

The Doomsday Argument provides a thought-provoking perspective on humanity's potential fate. It integrates probability, statistics, and philosophical principles, offering a statistical guess at our collective demise. While it is far from conclusive, it is certainly important in serving as a reminder of our finite earthly existence and the urgency to address the global threats that could precipitate our doom. Whatever else, the debate around the argument and our ultimate fate as humans will persist, sparking further exploration into this fascinating intersection of probability, philosophy, and existential prediction.

When Should We Stop Looking and Start Choosing? Exploring the Secretary Problem

WHEN TO STOP LOOKING AND START CHOOSING

The 'Secretary Problem' is a classic scenario in decision-making and probability theory. The 'Optimal Stopping Problem' or 'Secretary Problem', as it is often called, offers insights into the dilemma of when to stop looking and start choosing. Whether it is finding the right partner, hiring the best assistant, or identifying the ideal place to live, this mathematical problem delivers a powerful solution.

CHOOSING A CAR

Let's say that you have 20 used cars to choose from, offered to you in a random sequence. You have three minutes to evaluate each. Once you turn one down, there is no returning to it, such is the speed of turnover, but the silver lining is that you are guaranteed the sale of any vehicle you do select. If you come to the end of the line, you must accept whatever remains, even if it happens to be the least desirable. Your decision is guided solely by the relative merits of the vehicles on offer.

BALANCING BETWEEN TOO EARLY AND TOO LATE

There are two significant failures in your quest to find the best vehicle for you—stopping too early and stopping too late. If you stop too early, you might miss out on a better option. Conversely, if you stop too late, you risk passing over the best option while waiting for a better option that might not exist. So, how do you find the right balance?

INTRODUCING THE OPTIMAL STOPPING STRATEGY

Do you have a strategy that is better than random selection?

The Optimal Stopping Problem provides a solution. If there are three cars in the flash sale, optimal stopping strategy suggests rejecting the first option in order to gain more information about the relative merits of those available. If the second option turns out to be worse, you should wait, despite the risk of ending up with the third, which could potentially be the worst of the three. However, if the second option is better, you should accept it immediately, foregoing the possibility that the third might be a better match.

EXTENDING THE STRATEGY: FROM 4 TO 100

With four options, you should reject the first. Again, if the second is better than the first, take that. If not, and the third is better, take that. Otherwise, you must take the fourth and hope for the best. With a hundred options, you should inspect the first 37 and then choose the first after that which is better than the best of the first 37.

This strategy, often referred to as the 37% Rule, is based on the mathematical constant, e (Euler's number). The value of $1/e$ is approximately 0.36788% or 36.788%, which rounds up to 37%. Following this rule, you have a 37% chance of finding the best car by employing this strategy.

THE GROUNDWORK

When faced with a choice of n candidates for a job, the challenge lies in deciding when to stop the process of rejection and start the process of selection. The mathematical answer to this, as highlighted before, is to reject the first n/e candidates, where 'e' is the base of natural logarithms, approximately 2.7. So, if there are 100 choices, n/e becomes 100/2.7, which is about 37. This strategy effectively breaks the selection process into two phases: the assessment phase and the selection phase.

The resulting principle is, therefore, surprisingly straightforward: reject the first 37% of candidates to gather information about the quality of the pool, then select the next candidate who is better than anyone seen so far.

REAL-WORLD APPLICATIONS

While the Secretary Problem is a simplified and somewhat idealised situation, the 37% Rule can have valuable applications in real-world scenarios:

1. **Job Hiring:** Hiring managers can use the 37% rule as a strategic guideline during the candidate evaluation phase.
2. **Home Hunting:** The principle is also applicable as a heuristic when looking for a home to buy or rent, especially in a fast-moving msrket.
3. **Online Shopping:** This principle can also be useful when shopping online to streamline purchasing decisions. By reviewing a certain portion of available options before making a selection, shoppers can reduce the overwhelming array of choices and enhance their overall shopping efficiency and satisfaction.

CRITIQUES AND LIMITATIONS

While the 37% Rule provides a theoretically optimal solution to the Secretary Problem, it does have certain limitations:

1. **Idealised Assumptions:** The problem assumes that options are presented one at a time, in random order, and once rejected, they cannot be recalled.
2. **Risk of Missing Out:** Following the 37% Rule means you run the risk of the best option being rejected during the assessment phase.
3. **Difficult to Determine the Total Pool:** The problem assumes you know the total number of options upfront.
4. **Emotional Considerations:** The rule neglects emotional considerations, personal intuition, and human subjectivity.

ADAPTING THE RULE FOR UNCERTAINTY

The rule can be adapted if there is a probability that your selection of a range of options might opt out or be withdrawn. For example, if there is a 50% chance that the selection might opt out or be withdrawn after selecting it, then the 37% rule can be converted into a 25% rule, reflecting the added uncertainty. There is also a rule-of-thumb for when the aim is to select a good option, if not necessarily the best. Out of 100 options, for example, the square root rule suggests seeing the first ten (the square root of 100) and then selecting the first option of those remaining that is better than the best of those ten.

CONCLUSION: EXPLORATION AND EXPLOITATION

The Secretary Problem teaches us about the balance between exploration (gathering information) and exploitation (making a decision), offering a structured approach to navigating complex choices. Despite limitations, the Secretary Problem and the 37% Rule offer valuable insights into these trade-offs. It also provides a mathematically grounded approach to making complex decisions.

When Should We Take a Leap of Faith? Exploring Pascal's Wager

THE ORIGINS OF PASCAL'S WAGER

To understand the significance of Pascal's Wager in decision-making processes, we must first trace its roots. Blaise Pascal is known for his immense contributions to mathematics and probability theory. One of his notable contributions to philosophy and decision theory, however, was his articulation of what has come to be known as Pascal's Wager.

PASCAL'S WAGER: THE CRUX OF THE ARGUMENT

The wager posed by Pascal is simple yet profound. It can be paraphrased as follows: If God exists and you wager otherwise, the repercussions can be enormous. On the contrary, if God does not exist and you wager that he does, the implications are trivial in relative terms. Essentially, believing in God could lead to infinite rewards (eternal life in heaven), while the downside if he does not is comparatively inconsequential. Thus, Pascal urges you always to lean to the side of believing in God.

ADDRESSING THE 'MANY GODS' OBJECTION

The argument often raised against Pascal's Wager is the 'many gods' objection. Detractors argue that numerous characterisations of God are conceivable, including those that punish believers. However, this counterpoint presumes that all representations of a god are equally plausible, which is an assumption that may not hold.

For instance, the existence of a deity described by a major established religion with millions or even billions of adherents and millennia of theological development and intellectual underpinning could be perceived as vastly more plausible than a fledgling religion with relatively few adherents or consistent theology.

THE ROLE OF HUMAN BIASES AND FUTURE REWARDS

The ability to appreciate uncertainty and the value of future rewards too often gets overshadowed by human biases. Humans are predisposed to discount the future, focusing on immediate rewards and overlooking long-term consequences. This cognitive bias makes people prone to underestimate future risks and rewards. Pascal's Wager prompts us to consider future implications more seriously, offering a framework to factor in future gains or losses in decision-making.

PASCAL'S WAGER IN CONTEMPORARY CONTEXT: CLIMATE CHANGE AND NOAH'S LAW

The relevance of the thinking behind Pascal's Wager isn't confined to theological considerations. A parallel can be drawn, for example, between Pascal's Wager and the urgency to act against climate change. Even if there were only a slim chance of catastrophic climate disaster, the consequences of inaction, considering the potential existential harm, would be too high to ignore.

This approach to climate change action has been dubbed 'Noah's Law'. It reflects the sentiment of Pascal's Wager: if there's a chance an ark may be essential for survival, it's prudent to start building it now, regardless of how sunny the day might seem.

THE GUARDIAN PRINCIPLE

Building upon these concepts, I propose the introduction of a new ethical and operational guideline, which I call the 'Guardian Principle'. This principle extends the foundational ideas of Noah's Law and Pascal's Wager into a broader, more encompassing approach. It advocates for a stance of proactive stewardship over our planet and society, emphasising the importance of preemptive action in the face of potential existential threats, not limited to climate change but extending to all manner of such risks.

The Guardian Principle calls for an ethos of precaution and responsibility, urging humanity to act as guardians of its own future and the future of our shared environment. It suggests that in situations of significant uncertainty but potentially devastating outcomes, we should err on the side of caution and engage in preventative measures against a wide array of existential risks. In this way, we fulfil a collective duty to safeguard the well-being of current and future generations against all forms of irreversible harm.

By integrating the Guardian Principle into our global ethos, we expand the narrative from merely avoiding disaster to actively cultivating a safe, sustainable future. It's a call to not only build arks against impending floods but to seek to prevent the floods themselves, and to act more broadly as vigilant guardians against potential threats. It encourages us not just to react but to anticipate, mitigate, and ideally avert existential risks through foresight, innovation, and collective action.

In this light, the Guardian Principle does not just complement the logic behind Pascal's Wager and Noah's Law; it amplifies it. It reinforces the argument that inaction in the face of existential uncertainty is not an option. Instead, we are urged to embrace a more vigilant, proactive approach, turning existential anxiety into a catalyst for holistic and forward-thinking action. In this way, it is a call for a shift in perspective – from reactive measures to a stance that actively seeks to prevent, mitigate, and anticipate risks before they manifest. It's about building a legacy of sustainability, resilience, and foresight. It is a call to action that resonates with Pascal's Wager.

PASCAL'S MUGGING: A MODERN SPIN

A modern spin on Pascal's Wager, Pascal's Mugging, presents a scenario where a stranger promises a life-changing return on a relatively small sum, or to wield an existentially negative impact if they don't receive this sum. Even if the chance of the claim being true is infinitesimal, it might seem rational to hand over the sum, such is the scale of the reward or consequence compared to the outlay. It is a dilemma that underscores the need for a pragmatic balance between scepticism and action in the face of uncertainty.

How can we distinguish between genuinely plausible threats and those that, despite their potential for immense impact, are so improbable as to be practically disregarded? It is a question that is not merely academic but has real-world implications in risk management, policy-making, and personal decision-making.

In this way, a case can be argued for distinguishing between outcomes which are at least plausible, even if uncertain, and scenarios of vanishingly small probability. The problem, of course, is in determining which is which.

CONCLUSION: PASCAL'S LIGHTHOUSE

Pascal's Wager has assumed a new critical relevance in our times. With the stakes being higher than ever in terms of global existential risks, the urgency to revisit and appreciate the wager's lessons has heightened.

While it might seem counterintuitive to expend resources and energy to avert what may be perceived by at least some as small risks, Pascal's Wager prompts us to think otherwise. As the wager illuminates, the potential stakes of inaction—be it eternal damnation in a theological context or irreversible climate disaster in a worldly sense—may far outweigh the cost of preventive measures. As we steer through a world beset with systemic risks and uncertainties, Pascal's Wager, and the Guardian Principle it inspires, can serve as a lighthouse, guiding us away from the rocks and towards prudence, long-term thinking, and existential risk management.

When Should We Trust the Numbers? Exploring Benford's Law

BENFORD'S LAW: A STATISTICAL ANOMALY

Benford's Law reveals a surprising pattern in numerical data: lower digits, especially '1', appear more frequently as the leading digit in many real-world datasets than would be expected if all digits were equally likely.

This pattern, also known as the Newcomb–Benford Law, is a rule that goes against what most people would assume to be true about numbers.

Surprisingly, the probability of the leading digit being '1' is approximately 30%, instead of about 11%, which would be the case if all digits from 1 to 9 were equally probable. The number '2' would appear as the first digit about 17.6% of the time, '3' around 12.5%, and so forth, with the number '9' having a mere 4.6% chance of being the first digit.

A fascinating aspect of Benford's Law is its applicability to numerous datasets that reflect natural and societal phenomena, ranging from the populations of cities and countries to the heights of mountains and even utility bills.

Suppose, for example, you select numbers from everyday data like electricity bills, stock prices, or river lengths. You'll notice that numbers starting with '1' occur more often than those starting with '9'. In fact, you'll find that roughly 30% of these numbers will start with a 1.

One might wonder why this happens. The answer is rooted in the multiplicative processes that often generate the numbers we see in real life. For a number to grow from 1 to 2, it needs to increase by 100%. But to go from 2 to 3, the increase is only 50%. As a result, numbers spend more time with a leading digit of 1 before moving on to higher first digits, which leads to the distribution observed in Benford's Law. To look at it another way, if you start with £1 and it grows by 10% each day, the leading digit will stay as '1' for a longer duration than '2', and '2' will hold the top spot longer than '3', and so on.

CONDITIONS FOR BENFORD'S LAW

There are specific conditions that must be satisfied for Benford's Law to apply. One such condition is that the numbers should be of the same nature—mixing areas of lakes with employee numbers, for instance, won't adhere to the law.

The numbers also shouldn't have any artificial caps or boundaries. This excludes data such as house numbers or the price ranges of wines in a supermarket.

Moreover, the numbers shouldn't be designated by a specific numbering system, like postcodes or telephone numbers, nor should there be any significant spikes around particular numbers.

THE ORIGINS OF BENFORD'S LAW

This unusual phenomenon can be traced back to a note in the American Journal of Mathematics in 1881 by astronomer Simon Newcomb. Newcomb noticed that the pages of logarithms used to perform calculations were more worn for certain leading digits than others. Numbers starting with 1, for example, were used much more frequently than those beginning with 9.

However, it was physicist Frank Benford who provided rigorous empirical support for this distribution. His 1938 paper, 'The Law of Anomalous Numbers', examined 20,229 sets of numbers, ranging from baseball statistics to areas of rivers and numbers in magazine articles. He confirmed that approximately 30% of these numbers started with 1. The chances of the leading digit being '2' were found to be 17.6%, and for '9', a mere 4.6%.

IMPLICATIONS FOR FRAUD DETECTION

Interestingly, Benford's Law has found crucial applications in the detection of fraud, particularly in accounting. If declared returns significantly deviate from the Benford distribution, it serves as a red flag for those investigating fraudulent activities.

SCALE INVARIANCE: A UNIQUE PROPERTY

A truly universal law that governs the digits of numbers describing natural phenomena should operate independently of the units used. This property, known as scale invariance, holds true for Benford's Law. Regardless of the units of measurement, whether they are inches, yards, centimetres, or metres, the overall distribution of numbers would still maintain the same pattern.

THE MATHEMATICS BEHIND BENFORD'S LAW

Benford discovered that the probability that a number starts with n is equal to log $(n+1)-\log (n)$, to base 10, so that the probability that a number starts with 2, say, is equal to $\log_{10} (2+1)-\log_{10} 2=0.1761$, or 17.61%.

PREDICTING PROPORTIONS OF DIGITS WITH BENFORD'S LAW

In addition to the first digit, Benford's Law can also predict the proportions of digits in the second number, the third number, and so forth.

CONNECTING WITH NATURE

Benford's Law is all around us, even within the Fibonacci numbers. To explain, in the Fibonacci sequence each number is the sum of the two preceding ones, resulting in a series like 1, 1, 2, 3, 5, 8, 13, and so forth. This sequence is famously associated with the

Golden Ratio (approximately 1.618), which frequently appears in art, design, and even in natural patterns like the spiral pattern of sunflower seeds. The leading digits of these numbers, when they grow large enough, conform to Benford's Law. This intersection between Benford's Law and the Fibonacci sequence further accentuates the pervasive and mysterious nature of mathematical patterns in our universe.

CONCLUSION: THE BEAUTY OF BENFORD

Benford's Law serves as a testament to the fact that natural and other types of data are governed by underlying patterns, often too subtle for the naked eye but unmistakable in their mathematical signatures. The inherent bias towards smaller leading digits reflects the multiplicatively growing nature of many phenomena in our universe, an echo of the processes that shape everything from economics to geology.

While the conditions for Benford's Law provide boundaries for its application, they also highlight the importance of context in statistical analysis. This law does not claim universal application; rather, it thrives within the appropriate datasets—those untainted by human constructs or arbitrary limits.

In understanding Benford's Law, we find more than just an anomaly. We uncover a bridge between the abstract world of numbers and the tangible reality we measure, count, and analyse. Whether it's in unravelling potential fraud, examining natural occurrences, or exploring the mathematical patterns in art and nature, Benford's Law remains a powerful ally to the inquisitive mind.

When Should We Second Guess? Exploring a Number Dilemma

THE NUMBER DILEMMA

In the Number Dilemma, participants must choose a whole number between 0 and 100, aiming to get closest to two-thirds of the average number chosen by all participants. This scenario tests not only numerical reasoning but also understanding of human behaviour.

LEVEL 1 RATIONALITY: CHALLENGING THE AVERAGE

If you were to assume that the other participants would choose a random number within the given range, the average number chosen by everyone would be 50. Under this assumption, you might believe that choosing 33, the nearest integer to two-thirds of 50, would provide a high probability of winning. This initial strategy, known as Level 1 rationality, might appear intuitively logical.

LEVEL 2 RATIONALITY: ANTICIPATING THE AVERAGE OF THE AVERAGE

However, upon closer inspection, a new insight emerges. Since you reasoned that choosing 33 was a smart move, it is reasonable to assume that other participants will arrive at the same conclusion. Consequently, the average number chosen by all participants would shift towards 33. To maximise your chances of winning, you decide to adopt Level 2 rationality and choose a number lower than 33. In this case, 22 appears to be the optimal choice.

LEVEL 3 RATIONALITY: GOING DEEPER INTO ANTICIPATION

As you delve deeper into the rationality levels, a pattern begins to emerge. Just as you contemplated that others might select 22, they too will likely adopt the same line of reasoning. To outsmart them, you employ Level 3 rationality and opt for the number 15. The idea is to anticipate the choices of others and select a number that is two-thirds of the average they might choose.

LEVELS OF RATIONALITY

In summary, the levels of rationality illustrate the iterative process of outthinking others.

APPROACHING ZERO: THE ULTIMATE RATIONAL CHOICE

As you progress through each level of rationality, however, you cannot help but notice a concerning trend. As rationality levels increase, choices converge towards zero, posing a paradox: Is zero really the most rational choice when considering human diversity in decision-making? Deep down, you begin to question the effectiveness of choosing zero.

ACCOUNTING FOR DIFFERENT LEVELS OF RATIONALITY

Your uncertainty arises from the realisation that not all participants are likely to think or reason in the same way. Variations in human rationality mean that some may choose randomly or with less strategic depth, affecting the overall average and optimal choice.

THE WINNING NUMBER

In a practical application of this dilemma involving Financial Times readers, the winning number was 13, showcasing the unpredictability of collective rationality.

KEYNESIAN BEAUTY CONTEST IN FINANCIAL MARKETS

The economist John Maynard Keynes encapsulated the essence of the dilemma in his work, 'General Theory of Employment, Interest, and Money'. He likened professional investment to a newspaper competition where participants must select the prettiest faces from a selection of photographs. The prize is awarded to the competitor whose choices align most closely with the average choice of all participants.

OUTGUESSING THE BEST GUESSES

Keynes emphasised that participants should not merely choose what they believe to be the prettiest faces according to their own judgment or average public opinion. Instead, they should anticipate what average opinion expects the average opinion to be. In essence, winning the competition relies on outguessing the best guesses of others—a strategy referred to as super-rationality. Just as in the Number Dilemma, the Keynesian Beauty Contest involves predicting others' predictions, a strategy crucial in financial markets and investment decisions.

DISCOVERING THE HIDDEN OPPORTUNITIES

In the context of this so-called Keynesian Beauty Contest, the concept of super-rationality holds tremendous significance. This strategy involves outthinking the crowd's average opinion, a concept that can reveal overlooked opportunities in various contexts. By transcending the common line of reasoning and adopting a super-rational approach, individuals can unveil hidden possibilities and potentially reap rewards. While these concepts offer intriguing insights, their practical application is complex due to the unpredictable nature of human decision-making and diverse levels of rationality.

CONCLUSION: EMBRACING SUPER-RATIONALITY

The Keynesian Beauty Contest serves as a captivating thought experiment that challenges traditional notions of rational decision-making. It showcases the complexities of human behaviour and highlights the importance of anticipating the actions of others. By embracing the concept of super-rationality and outguessing the best guesses of the crowd, individuals can navigate these intricacies and increase their chances of success.

When Should We Expect to Be Late?
Exploring an Observer Selection Effect

THE SLOWER LANE PARADOX

Is the line next to you at the airport check-in or the supermarket check-out always quicker than the one you are in? Is the traffic in the neighbouring lane always moving a bit more quickly than your lane? We've all experienced it. Or does it just seem that way?

THE ILLUSION OF THE SLOWER LANE

One explanation for the perception of always being in the slower lane can be attributed to basic human psychology. Our tendency to notice and remember the times when we're overtaken, while quickly forgetting the moments we overtake others, may play a role in this feeling. Or might it be an illusion caused by our tendency to glance over at the neighbouring lane more often when we are progressing slowly rather than quickly? Additionally, our focus tends to be more forward-looking, causing vehicles we overtake to quickly fade from our memory while those remaining in front continue to torment us.

The question then arises: Is this perception all an illusion or is there a real and fundamental phenomenon at play? Philosopher Nick Bostrom suggests that the effect is real and is the consequence of an observer selection effect. It is not just a trick of the mind.

THE SELECTION EFFECT

To understand why we might frequently find ourselves in the slower lane, let's consider an example of fish in a pond. If we catch 60 fish, all of which are more than six inches long, does this evidence support the hypothesis that all the fish in the pond are longer than six inches?

The answer depends on whether our net is capable of catching fish smaller than six inches. If the holes in the net allow smaller fish to pass through, our sample of fish would be biased towards the larger ones. This is known as a selection effect or an observation bias.

Now, just as a fisherman's net biased towards larger fish can misrepresent the pond's population, our position in a slower lane biases our perception of overall speed and flow.

RANDOMLY SELECTED OBSERVERS

When considering whether we are more often in the slower lane, it is crucial to ask: 'For a randomly selected person, are the people or vehicles in the next line or lane actually moving faster?' We need to view ourselves as random observers and think about the implications of this perspective for our observations.

An apparent reason why we might find ourselves in a slower lane is the higher number of vehicles present compared to the neighbouring lane. Cars travelling at higher speeds are generally more spread out than slower cars, so a given stretch of road is likely to have more cars in the slower lane. Consequently, the average driver will spend more time in the slower lane or lanes. This phenomenon is known as an observer selection effect, where observers should reason as if they were randomly selected from the entire set of observers.

THE VIEWPOINT OF THE MAJORITY

To put it simply, if we perceive our present observation as a random sample from all observations made by all relevant observers, the probability is that our observation will align with the perspective of most drivers, and these are typically in the slower-moving lane. Because of this observer effect, a randomly selected driver will not only seem to be in the slower lane, but will actually be in the slower lane.

In other words, when we view ourselves as part of a larger group of observers, we realise that being in a slower lane is more than perception; it's a statistical likelihood.

For instance, if there are 20 observers in the slower lane and 10 in the equivalent section of the other faster lane, there is a 2/3 chance that we are in the slower lane.

CONCLUSION: EMBRACING THE REALITY

So the next time we think that the other lane is faster, we should be aware that it very probably is. Our perception aligns with the reality that the slower lane tends to have more observers, leading to a higher likelihood of finding ourselves in it.

Understanding the Slower Lane Paradox isn't just about traffic or queues, though. It's a lesson in perspective and probability, reminding us that our individual experiences often reflect broader statistical realities.

When Should We Expect the Evidence to Lie? Exploring Berkson's Paradox

BERKSON'S PARADOX

Berkson's Paradox, also known as Collider bias, is a statistical phenomenon where unrelated factors appear linked due to selective sampling. It's a common pitfall in data analysis that can lead to misleading conclusions.

AN ILLUSTRATIVE EXAMPLE: COLLEGE ADMISSION

To better understand this statistical phenomenon, consider an example involving a prestigious college. This institution only accepts students who demonstrate excellence in either music or sports.

Now, in the wide pool of all students, there is no direct connection between musical talent and sports proficiency. However, the college's unique selection criterion gives rise to a strange observation. Once we only consider the admitted students, it seems that those with exceptional musical talents don't generally excel in sports, and similarly, those who are sports stars generally don't show much musical ability.

This apparent negative correlation is simply an outcome of the way the college handpicks its students. The selection process inadvertently creates a group that displays a misleading negative association between musical and sporting abilities.

EXPLORING THE RELATIONSHIP

To understand what is going on, imagine that the college only admits students who score 100 in either sports or music, but their talent in the other subject is simply representative of the wider student average in that subject. So among the admitted students, the average score for sports among musically talented students is just 50, contrasting with 100 for music, and vice versa. This results in the illusion of a negative correlation between sports and musical abilities, where no such relationship exists in the wider student population.

BERKSON'S PARADOX IN EVERYDAY LIFE: THE BOOKS AND MOVIES CONUNDRUM

Berkson's Paradox can creep into our daily lives, sometimes in unexpected ways. A common example revolves around the adaptation of books into movies. Have you ever noticed how often an excellent book seems to become a disappointing movie?

This observation could be a manifestation of Berkson's Paradox. By focusing on the cases where good books are made into bad movies, we might miss the instances where good books are made into good movies, or bad books into bad movies, or even mediocre books into mediocre movies. These overlooked instances are crucial in determining the true nature of the correlation.

The key point is that there might be many successful adaptations of good books, but these are not as memorable or discussed as the disappointing ones. This leads to a perception bias. To accurately determine if there's a negative correlation, we would need to systematically analyse a large and representative sample of book-to-movie adaptations, considering all combinations of book and movie quality. Otherwise, we may be inadvertently selecting a skewed sample. The selective memory here mirrors the selection bias of the college in our previous example.

THE ATTRACTIVENESS MISCONCEPTION: AN APPLICATION OF BERKSON'S PARADOX

A slightly tongue in cheek example of Berkson's Paradox relates to attractiveness and demeanour. It's not uncommon, so the narrative goes, to observe that attractive people tend to have an unpleasant attitude ('handsome men are jerks', in the words of statistician Jordan Ellenberg). Ellenberg explains that this perception can be explained if we tend to avoid or ignore individuals who are both unattractive and unpleasant. This leaves us mixing with plenty of attractive, unpleasant people and plenty of pleasant, unattractive people, but very few unpleasant, unattractive people. In our circles, therefore, we see a negative association between attractiveness and being pleasant even if no such association might exist in the wider population.

BERKSON'S PARADOX: A SOURCE OF BIAS IN SAMPLE CONSTRUCTION

More generally, Berkson's Paradox serves as a reminder of the potential pitfalls of sample construction. When we focus on certain aspects of interest and ignore others, we risk creating a biased sample. Such a sample may indicate a negative correlation where none exists in the whole population.

BERKSON'S PARADOX IN COVID-19 RESEARCH

A study published in 2020 suggested that smokers were less likely to be hospitalised due to COVID-19. Does this mean that smoking protects from Covid? A more likely explanation is that the study was an example of Berkson's Paradox. Hospitalised patients were not a random sample of the population. They consisted of older individuals, frail individuals, smokers, and those with COVID-19.

To illustrate the problem, assume for simplicity that patients are admitted either for smoking-related illnesses or COVID-19. Covid tests in hospitals would likely now show lower infection rates among smokers because they're already hospitalised for smoking-related illnesses. Non-smokers would on the other be admitted because they have Covid. This creates the appearance of a negative association between smoking and rates of COVID-19 infection even though no such negative association exists in the wider population.

HOW TO AVOID THE BIAS

When analysing data, it's crucial to consider the whole population and avoid focusing on a selective sample that could distort the real picture.

CONCLUSION: INTERPRETING DATA

Berkson's Paradox teaches us the importance of comprehensive and unbiased data analysis. It reminds us that the way we select and interpret data can shape our understanding of the world. Understanding and recognising this paradox is crucial in various fields, including but not limited to social sciences, medical research, and data analysis.

When Should We Expect the Putt to Drop? Exploring Benchmarks and Behaviour

THE PROBLEM OF LOSS AVERSION

Loss aversion, a key concept in behavioural economics, explains our tendency to prefer avoiding losses over acquiring equivalent gains. This principle is illustrated in diverse settings, from biblical parables to professional sports.

LOSS AVERSION IN THE VINEYARD

The Gospel of Matthew in the New Testament of the Bible relates the Parable of the Workers in the Vineyard. Here's a breakdown of the parable:

Setting: A landowner goes out early in the morning to hire workers for his vineyard.

1. **Early Morning Hiring:** The landowner hires the first group of workers at daybreak, agreeing to pay them a denarius for the day's work, which was a typical day's wage at the time.
2. **Subsequent Hirings:** Throughout the day, the landowner hires additional groups of workers. He goes out again at the third hour (around 9 a.m.), the sixth hour (noon), the ninth hour (3 p.m.), and even the 11th hour (5 p.m.) to hire more workers. With these groups, he doesn't specify an exact wage, but promises to pay them 'whatever is right'.
3. **End of the Day—Payment Time:** When evening comes, the workers line up to receive their wages, starting with those who were hired last. To everyone's surprise, the workers hired at the 11th hour receive a full day's wage (a denarius).
4. **Discontent among the First Hired:** Seeing this, the workers who were hired first expect to receive more than their agreed-upon denarius. However, they too receive the same wage. This leads to grumbling against the landowner. They feel it's unfair that they worked the whole day, bearing the heat, and yet received the same as those who worked just an hour.
5. **Landowner's Response:** The landowner points out that he did not cheat the first workers; he paid them the agreed-upon wage.

Reference Points and Relative Outcomes: The workers who commenced early had a clear reference point—a denarius for a day's toil. Despite receiving the agreed wage, they perceived receiving the same pay as those who worked less as a loss. Observing others receiving equivalent wages for fewer hours of work framed their wages in this way. This episode underlines the significance of reference points in human decisions, emphasising the relational aspect of outcome evaluations, surpassing absolutes. The dissatisfaction emanating from the early workers is a classic example of loss aversion. It is a central feature of modern Prospect Theory, albeit some 2,000 years ahead of its time.

BENCHMARKS AND BEHAVIOUR

An intriguing study on the behaviour of New York City cab drivers focused on how they decided the duration of their work shifts in relation to their daily earnings. Contrary to the expectations set by traditional economic theories, which suggest workers would maximise their hours on days with higher demand (and thus potential earnings), the study discovered that cab drivers tended to work fewer hours on days where they were earning more per hour and worked longer hours on less lucrative days. This behaviour aligns more closely with the concept of setting earning targets—drivers tended to end their shifts once they reached a certain income goal, regardless of how quickly they achieved it. This finding, led by Colin Camerer and Linda Babcock, challenged the rational agent model in economics, suggesting that real-world decisions are influenced more by psychological factors and personal benchmarks than traditional economics would expect.

BENCHMARKS ON THE GOLF COURSE

In the game of professional golf, the reference point takes on a very different aspect. This time the problem does not arise because those playing for just a couple of holes earn the same as those who complete four rounds. Instead, the problem arises from the so-called 'Par' assigned to each hole, which is a benchmark indicating the number of strokes that a skilled golfer, typically a 'scratch golfer' – who plays without any handicap – is expected to take to complete that particular hole under standard playing conditions. While the total number of shots should be the player's real concern in most competitions, regardless of the assigned 'par' scores, the fear of failing to achieve par on individual holes may trigger the influence of loss aversion. In this case, the aversion is to underperforming expectations.

FINDING EVIDENCE OF LOSS AVERSION

Contrary to what might seem rational, analysis of more than 2.5 million putts, with detailed measurements of initial and final ball placement, reveals that professional golfers are indeed significantly influenced by the artificial par reference point. Specifically, it can be shown from the data that golfers are less accurate when putting to score better than par on a hole than when aiming for par. The data suggests, as such, that professional golfers exert more effort to avoid missing par on a hole than in scoring better than par. But why? Their only true objective in most competitions should be to minimise the number of shots taken, regardless of the 'par' assigned to each hole.

A paper published by Devon Pope and Maurice Schweitzer in the American Economic Review in 2011 examines a range of possible explanations, systematically eliminating them one by one until the true cause becomes evident. They conclude that golfers view par as their 'reference' score. Therefore, a missed par putt is perceived by golfers, perhaps subconsciously, as a more significant loss than a missed 'birdie' putt, i.e. one shot better than par. The reality is that par is an artificial construct—all that really matters in strokeplay competitions is the total number of shots taken. This implicit mental bias, however, leads to irrational behaviour during the game, with golfers unable to adjust their strategy even when made aware of this bias.

Interestingly, the researchers also observed a tendency for equivalent 'birdie' putts to come up slightly short of the hole in comparison to par putts, further confirming the hypothesis that a fear of a loss to par impacts the players' putting strategies.

EXPLORING ALTERNATIVE EXPLANATIONS

Despite the compelling evidence for loss aversion, alternative explanations were considered. The possibility that birdie putts might originate from more precarious positions, for example, was explored. However, comprehensive data and rigorous analysis ruled out competing theories.

DYNAMICS OF LOSS AVERSION ACROSS TOURNAMENT ROUNDS

Interestingly, the accuracy gap between par and birdie putts varied across the tournament rounds. It was largest during the initial round and decreased significantly by the fourth round. This fluctuation suggests that the influence of loss aversion and the salience of the par reference point are neither automatic nor immutable and may be affected by factors such as competitor scores later in the tournament.

IMPLICATIONS: BENEFICIAL KNOWLEDGE FOR FORECASTERS AND PSYCHOLOGISTS

This unique insight into professional golfer behaviour has profound implications. It provides valuable information for sports forecasters or even those betting on the match in-play. Moreover, it serves as critical knowledge for sports psychologists working with professional golfers. If these psychologists could find a way to subtly reframe a golfer's perception of a birdie putt, they could significantly improve a golfer's performance and earnings over time.

CONCLUSION: PERSISTENCE OF BENCHMARKS AND LOSS AVERSION

This analysis demonstrates that even in a high-stakes, competitive market setting, loss aversion persists among experienced agents. Even top-performing golfers in the study displayed signs of this enduring bias, highlighting its pervasive influence in decision-making scenarios.

The concept has, of course, much broader significance than in competitive sports or even at the workplace. It has been shown to exist in so many of our personal choices and perceptions. By considering how we respond in our own lives to artificial benchmarks and reference points, it has the potential to significantly improve our everyday decisions and actions for the better.

A Question of Chance

4

When Should We Count Cards? Gaining an Edge

CARD COUNTING: A WINNING STRATEGY IN BLACKJACK

In 1962, Ed Thorp introduced a strategy that would forever change the landscape of blackjack: card counting. His book, *Beat the Dealer: A Winning Strategy for the Game of Twenty-One*, presented a system based on probability theory that allowed players to gain an advantage over the house. Since then, card counting has become a topic of fascination for blackjack players worldwide.

UNDERSTANDING THE BASICS OF BLACKJACK

To grasp the significance of card counting, it's essential to understand the fundamentals of blackjack. The basic objective of the game is simple: players aim to draw cards that beat the dealer's hand without exceeding a total of 21. While basic strategy provides players with a foundation for optimal gameplay, card counting takes it a step further by incorporating the knowledge of which cards have already been dealt.

THE CONCEPT OF CARD COUNTING

Card counting revolves around the concept that certain cards have a different impact on the game's outcome than others. By using a system to estimate the proportion of high

DOI: 10.1201/9781003402862-4

and low cards still in the deck, the technique allows players to adjust their betting and playing decisions based on the remaining composition of the deck.

POPULAR CARD COUNTING SYSTEMS

Several card counting systems have been developed over the years, each with its own approach to assigning values to the cards. Here are a few notable examples:

1. **Hi-Lo Count:** The Hi-Lo Count is one of the simplest and most popular card counting systems. It assigns a tag of +1 to low cards (2–6), a tag of 0 to neutral cards (7–9), and a tag of –1 to high cards (10-Ace). By maintaining a running count based on these tags, players can assess the overall composition of the remaining deck.
2. **KO Count:** The Knock-Out (KO) Count is another popular system. In this method, all 7s, 8s, and 9s are assigned a tag of +1, while 10s through Aces are assigned a tag of –1. The remaining cards are considered neutral (tag 0).
3. **Hi-Opt Systems:** Hi-Opt systems, such as the Hi-Opt I and Hi-Opt II, aim to provide a more accurate assessment of the deck's composition by considering more card values.
4. **Zen Count:** The Zen Count system is known for its precision in tracking the deck's composition. It assigns a variety of values to different cards, creating a more detailed count. This system, while more complex than the other systems, can offer a greater edge to skilled players.

Additional Considerations: It's crucial to understand that these systems vary in complexity and suitability for different players. Advanced systems like the Zen Count may offer more accuracy, but they require more practice and skill. Additionally, systems may require converting the 'running count' into a 'true count' by accounting for the number of decks remaining in the shoe. This adjustment helps in accurately determining the player's edge.

MAKING INFORMED DECISIONS

By monitoring the running count and employing the chosen card counting system, players can make in-running staking decisions. When the count indicates an abundance of high cards in the remaining deck or decks, this is generally good for the player, bad for the house. In this case, players may choose to increase the size of their bets. Conversely, when the count indicates a higher proportion of low cards remaining in the deck, players may opt for smaller bets and more conservative gameplay.

CHALLENGES AND COUNTERMEASURES

Casinos are well aware of card counting strategies and have implemented various countermeasures to detect and deter such activities. They employ techniques such as automatic shuffling machines, frequent deck changes, and trained personnel to identify suspected card counters. Consequently, players who employ card counting techniques also employ camouflage methods to avoid detection. This involves blending in with other players, varying bet sizes, acting like a casual player, and avoiding suspicious behaviour.

THE EVOLUTION OF CARD COUNTING

Over the years, card counting has evolved alongside advancements in technology and changes in casino practices. The rise of online blackjack games and continuous shuffling machines (CSMs) has posed new challenges for card counters. Online casinos employ random number generators (RNGs), making it impossible to track specific cards. CSMs continuously shuffle the cards, eliminating any opportunity to gain an advantage through card counting.

CONCLUSION: BEATING THE ODDS

Card counting revolutionised the game of blackjack by providing players with a mathematical strategy to gain an edge over the house. However, it requires skill and practice to implement while evading detection. Still, card counting remains a challenging yet fascinating aspect of blackjack gameplay, and players can in principle adapt their techniques to the countermeasures employed by casinos. It continues to captivate players who seek to test their skills and beat the odds at the blackjack table.

When Should We Up the Stake?
Exploring the Kelly Criterion

TAKING ADVANTAGE OF THE ODDS

One of the most critical aspects of any betting strategy is determining the size of the bet when we believe the odds are in our favour. The answer to this pressing question was formalised by John L. Kelly, Jr., an engineer at Bell Labs by profession, as well as a hobbyist daredevil pilot and recreational gunslinger. His methodology, known as the Kelly Criterion, is a mathematical formula designed to establish the optimal bet size when we have an advantage, that is, when the odds favour us.

However, having the advantage doesn't guarantee a successful outcome. Irrespective of our edge, excessive betting can lead to large losses and, in worst-case scenarios, to bankruptcy. This is where the Kelly Criterion comes into play. It takes account of both the size of our edge and the potentially damaging impact of volatility.

UNDERSTANDING THE KELLY CRITERION

The Kelly Criterion is essentially a mathematical formula that calculates the optimal amount to bet or invest when the odds are in our favour. The fundamental principle underpinning the Kelly Criterion is that the amount of capital wagered should be related to our advantage at the available odds. It emphasises the relationship between the size of our bet and our perceived edge.

Consider this simple illustration: Suppose you have a coin where the probability of getting a head (winning) is expected to be equal to the probability of a tail (losing). Now, suppose you have secret information that the next coin toss will certainly be heads. In this case, you have a 100% edge. According to the strict Kelly Criterion, you should bet your entire capital because you're guaranteed to win. In real-life applications, even with very high confidence, betting one's entire capital is risky, however, due to the possibility of unforeseen factors. This is more a thought experiment than a practical recommendation.

In any case, in most scenarios the outcomes are not binary, and the probability of winning is rarely 100%, even in theory.

Let's consider a different situation: You're still betting on the coin toss, but this time your secret information gives you a 60% chance of landing heads and a 40% chance of tails. Your edge is now 20%, and a very basic Kelly strategy is to stake 20% of your capital.

This example reflects the core concept of the Kelly Criterion. It's not only about gauging when you have the advantage—it's also important to understand precisely how much to stake when you do. In theory, this sounds simple, but in practice, accurately identifying that advantage can be complex.

On a broader scale, the Kelly Criterion can be employed in various fields beyond betting, such as investing and trading, to determine the optimal size of a series of bets or investments. Its aim is to maximise the exponential growth of the bettor's or investor's wealth over the long term.

A strength of the Kelly Criterion is its flexibility. It allows you to adjust the proportion of the capital that you bet based on how strong your edge is.

It's important to note, however, that the Kelly Criterion assumes that the bettor can reinvest their winnings. This is crucial for the 'compounding' aspect of the strategy, which is what allows the wealth to grow faster than it would with other systems. This compound growth strategy is what differentiates Kelly betting from more static strategies, but also introduces higher volatility in the short term.

Ultimately, the Kelly Criterion offers a robust methodology for managing risk and maximising returns when the odds are in our favour. However, as with any strategy, understanding the core principles is just the beginning—it's the accurate identification of the edge and the consistent application of the strategy that's critical for long-term success. Misestimations can lead to over-betting and significant losses.

APPLYING THE KELLY CRITERION

The application of the Kelly Criterion can have profound implications for various fields beyond gambling, such as investing and trading. The crucial component is to understand that the Kelly Criterion isn't just about betting when we have an edge; it's about calculating the precise amount to bet to maximise compounded return over time.

This is where the Kelly formula can come into play:

$$F = Pw - (Pl/W)$$

where

- F is the Kelly criterion fraction of capital to bet,
- W is the amount won per amount wagered (i.e. win size, net of the stake, divided by loss size),
- Pw is the probability of winning, and
- Pl is the probability of losing.

When we apply this formula, we calculate the optimal fraction of our capital to bet, given our probability of winning (Pw), our probability of losing (Pl), and our potential return (W).

Consider a simple example: Suppose we have an even-money bet, i.e. the amount you stand to win, net of the stake, is the same as the amount you risk. In this scenario, the value of W is 1. If our chance of winning is 60% and our chance of losing is 40%, substituting these values into the simplified Kelly formula ($F = Pw - Pl$) gives us $F = 0.60 - 0.40 = 0.20$ or 20%. This means that in order to maximise our long-term return, we should bet 20% of our capital.

Let's consider a slightly more complicated scenario: Suppose we have a bet where we stand to win double the amount we risk, i.e. $W = 2$, and the probability of winning and losing is both 50%. Substituting these values into the original Kelly formula gives us $F = 0.50 - (0.50/2) = 0.50 - 0.25 = 0.25$ or 25%. This means we should bet 25% of our working capital to maximise our long-term return.

The Kelly Criterion is designed to ensure that you never go bankrupt because the recommended bet size decreases as your capital decreases. However, this doesn't mean you can't lose money. The Kelly Criterion maximises long-term growth rather than short-term returns. This means that there will be times when you lose money, but over the long run, you should come out ahead.

It's crucial to remember that the Kelly Criterion assumes you know the true probabilities of the outcomes, which is often not the case. In practice, we're often working with estimated probabilities, which means there's a risk that we could, for example, overestimate our edge and bet too much. Therefore, many investors and bettors use a fraction of the Kelly Criterion (betting a fixed fraction of the amount recommended by Kelly) to reduce their risk.

Lastly, while the Kelly Criterion offers a mathematical approach to betting and investing, it doesn't account for the emotional aspect of risking money. Remember that the goal is not just to maximise returns, but also to sleep well at night.

POTENTIAL RISKS AND LIMITATIONS

While the Kelly Criterion can be an effective strategy for maximising the growth of capital in the long run, it is not without its potential risks and limitations. These should be understood and considered before applying the formula.

Estimation Errors

The effectiveness of the Kelly Criterion hinges on the accuracy of the probabilities used in the calculation. An overestimation of the probability of winning (Pw) can lead to excessive bet sizes and the risks associated with over-betting.

Minimum Bet Size

The Kelly Criterion presupposes that there is no minimum bet size, which is rarely the case in real-world scenarios, especially in investing and trading. In situations where a minimum bet size exists, the possibility of losing all of the capital becomes a reality if the amount falls below this threshold.

Risk Tolerance

The Kelly Criterion determines bet sizes purely based on mathematical calculations to maximise long-term growth. It does not take into account the individual bettor's or investor's risk tolerance. An aggressive bet size recommended by the Kelly Criterion may not be psychologically comfortable for some, causing stress and potentially leading to sub-optimal decisions.

Given these potential risks and limitations, it is common for many investors and bettors to use a fractional Kelly strategy, betting a fraction (like half or a third) of the amount recommended by the Kelly formula. This approach can help mitigate the risks associated with over-betting and inaccuracies in probability estimation while still providing the benefits of proportional betting and capital growth. However, even a fractional Kelly strategy should be tailored to individual circumstances, including risk tolerance and the ability to withstand potential losses.

CONCLUSION: TAKING ADVANTAGE OF OUR EDGE

The Kelly Criterion, devised by John L. Kelly, Jr., is a unique betting strategy that uses probability and potential payout to determine the optimal bet size when the odds are in our favour. The mathematical formula suggests betting a fraction of capital equivalent to the size of one's advantage. However, it's crucial to account for potential errors and uncertainties that can affect the real-world implementation of this strategy.

Uncertainty in the size of any actual edge at the odds and the potential for a bumpy ride due to volatility mean that we should always exercise caution. As a result, unless we're prepared for potential high volatility and have unwavering confidence in our judgment, adopting a fractional Kelly strategy might be the most prudent approach. This strategy allows us to stake a defined fraction of the recommended Kelly amount, reducing risk while still taking advantage of our edge.

When Should We Expect to Be Kicked by a Horse? Exploring the Poisson Distribution

A STATISTICAL TOOL

The Poisson distribution, inspired by the work of Siméon Denis Poisson, is a statistical concept that is particularly useful for helping us understand events that occur infrequently. It indicates the number of such events we can expect to occur in a fixed interval if we know the average rate at which they arrive. In simpler terms, if you want to predict how often something will happen over a certain period, and this event is infrequent, the Poisson distribution can be your go-to method for making this prediction.

This distribution finds practical applications in various fields, ranging from studying historical events to analysing everyday situations and even sports.

UNDERLYING ASSUMPTIONS OF THE POISSON DISTRIBUTION

The accuracy and applicability of the Poisson distribution hinge on several key assumptions:

1. **Independence of Events:** Each event must occur independently of the others. This means the occurrence of one event does not affect the probability of another event occurring.
2. **Constant Average Rate:** The events are expected to occur at a constant average rate. In other words, the average number of events per unit of time or space remains consistent throughout the period being considered.
3. **Random Occurrence:** The events occur randomly, without any predictable pattern or structure. This randomness is crucial for the Poisson model to provide accurate predictions.
4. **Discrete Events:** The events are distinct and countable. For instance, the number of emails received per day or the number of accidents at a particular intersection per month.

Understanding these assumptions is vital for correctly applying the Poisson distribution. It is most effective in situations where these conditions are met, such as modelling the number of meteor showers observed in a year, counting the number of times a rare bird is spotted in a forest, or predicting the number of cars passing through a toll booth in an hour.

It's also very useful in predicting how likely you are to be kicked by a horse next week! The next section explains.

PREDICTING RARE EVENTS: PRUSSIAN CAVALRY OFFICER DEATHS

Let's travel back in time to the 19th century, when the Poisson distribution was used to study a particular historical event. During this period, researchers were interested in understanding the number of Prussian cavalry officers who were kicked to death by horses in different Army regiments over a span of 20 years. This unfortunate occurrence was relatively rare, but was it random, or were there some underlying factors influencing their occurrence?

Enter Ladislaus Bortkiewicz, an economist and statistician. Bortkiewicz collected data from 14 corps over 20 years, which resulted in observations of yearly numbers of deaths per corps. Using the formula associated with the Poisson distribution, he was able to predict the number of such deaths in specific time intervals. These fitted quite closely to the observed data, indicating that the deaths were indeed random events, and nothing more mysterious or sinister.

This application of the Poisson distribution became a textbook example of real-world events that can be modelled as Poisson processes, which include radioactive decay, arrival of emails, number of phone calls received by a call centre, etc. The deaths of Prussian cavalry officers are an early example of a statistical study in the field of survival analysis.

WORLD WAR II BOMBING RAIDS

During the Second World War, a British statistician named R.D. Clarke used this method to study where the new V-1 'flying bombs' were falling in London. He wanted to figure out if the German military was successfully targeting specific areas or if the bombs were falling randomly. This was strategically important information. It was clear that the V-1s sometimes fell in clusters. The question was whether this could be expected from random chance or whether precision guidance was at play.

To find out, Clarke divided London into thousands of small, equal-sized areas. He assumed to start with that each area had the same small chance of being hit by a bomb. This situation was similar to playing a game many times where you 'win' only infrequently. Clarke's calculations showed that the number of bomb hits in each area matched what the Poisson distribution predicted for random hits. This meant that where the bombs fell seemed to be a product of chance, not because specific areas were targeted.

FROM HISTORY TO FOOTBALL: PREDICTING GOAL SCORING

In football, goals are a relatively infrequent event within the setting of a match, and so are suitable for the application of the Poisson distribution. This provides a simple and effective tool to examine and predict the likely incidence of goals in a match, based on historical data and average goal rates.

Consider, say, a match between two teams, one with an average goal rate of 1.6 goals per game and the other with an average goal rate of 1.2 goals per game. The Poisson distribution allows us to calculate the probabilities of various goal-scoring outcomes for this specific match.

For example, by examining the historical data and applying the Poisson distribution, analysts can estimate the probability of a goalless draw, a 1-1 draw, a win for either team, or any other scoreline based on the average goal rates of the teams involved.

More generally, the Poisson formula allows us to calculate the chance of observing a specific number of events of this kind when we know how often they usually occur on average. It considers the average rate and calculates the probability of obtaining the specific number we're interested in.

REAL-WORLD APPLICATIONS

The practical applications of the Poisson distribution extend far beyond historical events and sports analytics. This versatile statistical concept finds relevance in a wide range of modern real-world scenarios, helping us understand and analyse various phenomena. Let's explore some of its notable applications.

Homes Sold and Business Planning

Imagine you are a local estate agent. Understanding the number of homes you are likely to sell in a given time period is crucial for business planning and forecasting. The Poisson distribution provides a framework for estimating the probability of selling a specific number of homes per day, week, or any other timeframe based on historical data and average sales rates. This information helps in making informed decisions about marketing strategies, staffing, and resource allocation.

Disease Spread and Epidemiology

In the field of epidemiology, the Poisson distribution plays a vital role in understanding the spread of infectious diseases. By analysing historical data and considering the average rate of infection, researchers can utilise the Poisson distribution to estimate the likelihood of disease outbreaks and their progression.

Telecommunications and Network Traffic

The Poisson distribution finds application in the analysis of telecommunications systems and network traffic. By studying the arrival patterns of these events using the Poisson distribution, companies can anticipate network demand, allocate resources effectively, and ensure smooth and reliable communication services.

Quality Control and Manufacturing Processes

The Poisson distribution is also used in quality control, particularly in manufacturing settings. By analysing the number of defective products using the Poisson distribution, manufacturers can estimate the probability of observing a specific number of defects. This information helps them identify areas for improvement and enhance overall product quality.

Traffic Accidents and Road Safety

Another area where the Poisson distribution finds application is in analysing traffic accidents and road safety. By examining historical data on accidents, researchers can use the Poisson distribution to model accident rates based on factors such as location, time of day, and road conditions. This understanding helps in the development of targeted interventions to reduce accidents and improve road safety.

CONCLUSION: A POWERFUL TOOL FOR INFREQUENT EVENTS

The Poisson distribution is a valuable statistical tool that helps us understand and analyse events that happen infrequently but have an average rate of occurrence. It may seem complicated at first, but it allows us to make predictions and informed decisions based on probabilities. By using the principles of the Poisson distribution, we can gain insights into rare events and use that knowledge to improve various aspects of our lives.

When Should We Roll the Dice?
Exploring Some Games of Chance

UNDERSTANDING THE CHEVALIER'S DICE PROBLEM

Probability is the science of uncertainty, providing a way to measure the likelihood of events occurring. It can be viewed as a measure of relative frequency or as a degree of belief. In the context of gambling, understanding probability is crucial for making informed decisions and avoiding common pitfalls.

A famous problem, known as the Chevalier's Dice Problem, sheds light on the some of the intricacies of probability.

To understand the problem, it is essential to grasp some fundamental concepts of probability. Consider a single die roll—each outcome represents a possible event, such as rolling a 1, 2, 3, 4, 5, or 6. When rolling two dice, there are 36 possible outcomes (six outcomes for the first die multiplied by six outcomes for the second die).

THE FLAWED REASONING OF THE CHEVALIER

The Chevalier's Dice Problem originated from a gambling challenge offered by the Chevalier de Méré, a 17th-century French gambler. The Chevalier offered even money odds that he could roll at least one six in four rolls of a fair die.

The Chevalier's reasoning was based on the assumption that since the chance of rolling a six in a single die roll is 1/6, the probability of rolling a six in four rolls would be 4/6 or 2/3. However, this reasoning can be shown to lead to inconsistent results when extrapolated to more rolls.

The correct approach involves considering the independent nature of each throw of the die. The probability of a six in one go is 1/6, so the probability of not getting a six on that go is 5/6. To calculate the probability of not rolling a six in four throws, we multiply the probabilities: $(5/6) \times (5/6) \times (5/6) \times (5/6) = 625/1296$.

Therefore, the probability of at least one six in four attempts is obtained by subtracting the probability of not rolling a six in any of those four attempts from 1: $1 - (625/1,296) = 671/1,296 \approx 0.5177$, which is greater than 0.5.

Despite his faulty reasoning, the Chevalier still had an edge in this game by offering even money odds on an event with a probability of 51.77%.

THE CHEVALIER'S MISSTEP WITH THE MODIFIED GAME

Encouraged by his initial success, the Chevalier expanded the game to 24 rolls of a pair of dice, betting on the occurrence of at least one double-six. His reasoning followed the same flawed pattern: since the chance of rolling a double-six with two dice is 1/36, he believed the probability of at least one double-six in 24 rolls would be 24/36 or 2/3.

The correct probability calculation involved considering the independent nature of each dice roll. The probability of no double-six in one roll is 35/36. Therefore, the probability of no double-six in 24 rolls is (35/36) raised to the power of 24, which is approximately 0.5086.

Subtracting this value from 1 yields the probability of at least one double-six in 24 rolls: $1 - 0.5086 = 0.4914$, which is less than 0.5. Hence, the Chevalier's edge in this modified game was negative: $49.14\% - 50.86\% = -1.72\%$.

This outcome demonstrated that even if the odds seem favourable, incorrect reasoning can lead to erroneous conclusions. The Chevalier's faulty understanding of probability caused him to lose over time.

THE IMPORTANCE OF CORRECT PROBABILITY CALCULATION

These examples underscore the critical nature of accurate probability calculations in games of chance. While intuitive reasoning may seem convincing, it often leads to incorrect conclusions, as demonstrated by the Chevalier's bets. Understanding the true probability of events is essential for informed decision-making in gambling and many other contexts where risk and uncertainty play significant roles.

THE GAMBLER'S RUIN AND UNDERSTANDING FINITE EDGES

The Gambler's Ruin problem raises the complementary question of whether, in a gambling game, a player will eventually go bankrupt if playing for an extended period against an opponent with infinite funds, even if the player has an edge.

For instance, imagine a fair game where you and your opponent flip a coin, and the loser pays the winner £1. If you start with £20 and your opponent has £40, the probabilities of you and your opponent ending up with all the money can be calculated using the following formulas:

$P1 = n1/(n1 + n2); P2 = n2/(n1 + n2)$

Here, $n1$ represents the initial amount of money for player 1 (you) and $n2$ represents the initial amount for player 2 (your opponent). In this case, you have a 1/3 chance of winning the £60 (20/60), while your opponent has a 2/3 chance. However, even if you win this game, playing it repeatedly against various opponents or the same one with borrowed money will eventually lead to the loss of your betting bank. This holds true even when the odds are in your favour. This is an important lesson in risk management, emphasising the importance of not only the odds but also the size of one's bankroll relative to the stake sizes.

The Gambler's Ruin problem, as explored by Blaise Pascal, Pierre Fermat, and later mathematicians like Jacob Bernoulli, reveals the inherent risks of prolonged gambling, even with favourable odds.

PILOT ERROR: MISUNDERSTANDING CUMULATIVE PROBABILITY

In Len Deighton's novel 'Bomber', a statistical claim suggests that a World War II pilot with a 2% chance of being shot down on each mission is 'mathematically certain' to be shot down after 50 missions. This assertion is a classic example of misinterpreting cumulative probability. In reality, if a pilot has a 98% chance of surviving each mission, their probability of not being shot down after 50 missions is 0.98 to the power of 50 (0.98^{50})which is approximately 0.36, or 36%. Thus, their chance of being shot down over these 50 missions is 64% $(1 - 0.36)$, not 100%.

SURVIVORSHIP BIAS: THE CASE OF BULLET-RIDDEN PLANES

The concept of survivorship bias is vividly illustrated in the case of analysing planes returning from missions during World War II. Upon examining these planes for bullet holes, it was observed that most hits were on the wings, tail, and the body of the plane, with few on the engine. The initial, intuitive response might be to reinforce the areas with the most bullet holes. However, this would be a misinterpretation of the data.

The key realisation, identified by statistician Abraham Wald, was that the planes being analysed were those that survived and returned to base. The areas with fewer bullet holes, such as the engines, were likely critical to survival. Planes hit in these areas probably didn't make it back, hence the lack of data for these hits. This understanding exemplifies survivorship bias—focusing on survivors (or what's visible) can lead to incorrect conclusions about the whole population.

Wald's insight led to the reinforcement of seemingly less-hit areas like engines, contributing significantly to the survival of many pilots. His work in operational research during the war provided a critical perspective on interpreting data and making decisions under uncertainty.

CONCLUSION: DICE, ODDS, AND RUIN

The Chevalier's Dice Problem illustrates the importance of understanding probability in gambling scenarios. Probability theory, as developed through famed correspondence between Pascal and Fermat, has contributed to modern probability concepts and the understanding of risk involved in gambling.

The Gambler's Ruin is a kind of warning from the world of probability, telling us that in gambling, a slight edge is no guarantee of success. Imagine two gamblers, one with an edge over the other but with much less money to play with. Even if the first player is more likely to win each round, their thinner wallet means they could run out of money after a few bad games. In contrast, the player with the deep pockets can keep playing longer, until (given enough money) luck swings their way. This underlines the importance and impact of losing streaks in games of chance.

The wartime examples highlight the real-world importance of understanding probability and statistical concepts accurately. They serve as a reminder that intuition can often lead us astray. Correctly interpreting data, especially in high-stakes situations, can have life-saving implications.

When Should We Stake It All?
Exploring the Gambler's Dilemma

THE DILEMMA

When the stakes are high and time is not a luxury, finding a solution can be like gambling with fate. This was the scenario for Mike, needing £216 to settle an urgent debt, with only £108 in hand. The roulette wheel beckoned as a potential salvation, but what was the most effective strategy to double his money?

UNDERSTANDING THE ODDS IN ROULETTE

To fully grasp the situation that Mike finds himself in, it's crucial to examine the mechanics and probabilities of the game he's chosen as his lifeline: roulette. Specifically, we are considering a single-zero roulette wheel, a version of the game commonly found in European casinos.

Roulette consists of a spinning wheel and a small ball. The wheel is divided into 37 compartments or 'slots': numbers from 1 to 36 (randomly assigned as red or black) and a single zero slot. Bets can be placed on a single number, colour, or various combinations thereof.

In a single-zero roulette wheel, the player has a 1 in 37 chance of correctly predicting the outcome. This is because there are 37 slots in total: 36 numbers and the zero. So if you bet on a single number, the odds of the ball landing on that number are 1 in 37, or 36/1. The payout for such a bet, however, is 35/1. This discrepancy between the actual odds (36/1) and the payout odds (35/1) is where the house gains its edge. Every time a player wins, the house pays out less than the actual odds would dictate. In this way, the house earns a profit over time.

The 'house edge' is approximately 2.7%, a figure derived from the ratio of the single zero slot to the total number of slots (1/37). This constant advantage in favour of the casino is what makes the game fundamentally a game of negative expectation for players.

To understand the house edge in another way, consider this: if you were to place a £1 bet on each of the 37 slots, totalling £37, your return would be £36 (the £35 returned on the winning number plus the stake of £1). So for every £37 wagered, you would lose £1 using this strategy, which is approximately a 2.7% loss—exactly the house edge.

In conclusion, roulette, like all casino games, is a game of probabilities. And these probabilities, owing to the discrepancy between the actual odds and the payout odds, are slightly skewed in favour of the house. This fundamental understanding of the game's odds is pivotal when contemplating betting strategies, as we will see with the employment of 'bold' and 'timid' approaches.

THE BOLD STRATEGY: STAKING IT ALL

Mike's precarious situation leads him to contemplate a high-risk, high-reward approach known as the 'bold' strategy, which involves wagering all his available money at once. In this instance, he considers staking his entire £108 on the colour Red, a bet with almost a 50-50 chance, as the roulette wheel has 18 red slots out of 37 total slots.

To fully appreciate the audaciousness of this approach, it's essential to understand the mathematics behind it. When betting on a colour, there's a near-even split of potential outcomes: 18 red slots, 18 black slots, and the zero slot. Thus, the likelihood of the ball landing on a red slot is 18 out of 37, or roughly 48.6%. Consequently, with this single bet, he has about a 48.6% chance of doubling his money and obtaining the £216 he urgently needs.

However, it's important to note that this is a single-round probability. Unlike a 'timid' strategy, where multiple rounds are played, the bold strategy is a one-off scenario. Therefore, the 48.6% chance of winning must be interpreted as his overall chance of achieving his target sum. There are no second chances or opportunities to recoup losses; it's an all-or-nothing situation.

By putting all his money on one bet, he is maximising his return if that bet is successful. This is in contrast to a timid strategy, where the payout would be spread over multiple smaller bets, with the likelihood of achieving the target sum being significantly less.

But the bold strategy also comes with the highest level of risk. If the ball doesn't land on Red, Mike loses everything. His entire available funds are at stake, making the potential loss just as significant as the potential gain.

In conclusion, the bold strategy is a high-stakes, high-reward approach. It encapsulates the old saying, 'Who dares, wins', and, in this case, provides him the best chance of reaching his £216 target. Why is this so?

TIMID APPROACH: MULTIPLE SMALL BETS

As opposed to the bold strategy, he could consider dividing his available £108 into 18 separate bets of £6 each. These small, successive bets would be placed on a single number until he either depletes his funds or hits the winning number, which would yield a payout of 35 to 1, giving him the £216 he needs.

To fully understand the implications of this strategy, we need to analyse it in detail. The probability of winning a single number bet in roulette is 1 in 37, as there are 36 numbers and one zero. Hence, for each individual bet, John has a 1 in 37 chance of winning, or approximately 2.7%.

However, the timid strategy involves making multiple small bets, and so we must calculate the probability of these successive bets all losing. Since each individual bet has a 36 in 37 chance of losing, the probability that all 18 bets lose would be calculated as (36/37) to the power of 18, which equates to around 0.61, or 61%.

As such, the probability of him winning at least once using this timid strategy would be equal to 1 minus the losing probability. Hence, the chance of hitting the target £216 is 1 − 0.61, or 39%.

Interestingly, the timid strategy, although appearing less risky, significantly reduces Mike's chances of achieving his target sum compared to the bold approach. By spreading out his available funds across multiple bets, he lowers his exposure to loss in each individual game, but also decreases the likelihood of achieving his overall goal.

This strategy extends the length of play and the suspense, providing more instances of potential winning and losing. However, each bet also exposes Mike to the house edge, and therefore the risk of losses incrementally increases.

In this way, the timid approach offers more sustained engagement with the game but sacrifices the higher winning potential found in the bold approach.

THE POWER OF BOLD PLAY: TAKING A CALCULATED RISK

To look at it another way, consider a scenario where equal amounts are bet on red and black in each round. In most cases, the outcome will lead to breaking even, specifically 36 out of 37 times. However, when the ball lands on the single zero slot, the entire bank is lost. The more games played, the greater the chance of this happening.

By limiting the game to a single spin, the bold strategy minimises the number of times the house edge comes into play. Hence, playing fewer rounds decreases the likelihood of the house edge depleting the funds before reaching the target.

This strategy is not just about boldness in the face of risk, but more about understanding and working around the inherent disadvantage players face in casino games. By playing fewer games, you reduce the opportunities for the house edge to work against you.

CONCLUSION: THE INTUITION BEHIND BOLD PLAY

The intuition behind bold play in unfavourable games is grounded in a nuanced understanding of the mechanics of casino games and their built-in house edge. Bold play aims at striking hard and fast, capitalising on the relatively high chance of achieving the target sum in a single round, instead of facing the progressively increasing exposure associated with multiple rounds. In this sense, it's a calculated and strategic form of boldness.

When Should We Share the Pot? Exploring the Problem of Points

THE GENESIS OF THE PROBLEM

The Problem of Points, a pivotal moment in the history of probability, emerged from a gambling quandary addressed to Blaise Pascal by Chevalier de Méré in 1654. This problem centres around a fair method of dividing stakes in a game prematurely halted. Pascal, in turn, engaged Pierre de Fermat in this problem, setting the stage for a groundbreaking collaboration.

EARLY APPROACHES TO THE PROBLEM

Prior to Pascal and Fermat's contributions, notable figures like Luca Bartolomeo de Pacioli and Niccolo Fontana Tartaglia proposed their methods to resolve the issue. Pacioli suggested a division based on the rounds already won by each player. Tartaglia's solution was based on the lead of one player over another but also taking account of the number of rounds in the game. Neither method offered the correct general solution to the question.

THE PASCAL–FERMAT SOLUTION

Pascal and Fermat introduced an innovative approach that defied the prevailing intuition, focusing not on the current score but on the potential outcomes if the game had continued. This implies that an individual leading by 6-4 in a game to 10 should have the same winning odds as a player leading by 16-14 in a game to 20. Thus, the critical factor in their solution is not the number of rounds won by each player, but the number of rounds each player still needs to win.

FERMAT'S METHOD: ANALYSING POSSIBLE OUTCOMES

Fermat's method consists of examining the possible outcomes of a predefined number of additional rounds, assuming the game continues. For instance, in a game where Player 1 leads by 2-1, Fermat's method calculates the probability of each player winning based on the possible outcomes of four more games. Through this approach, a proportionate division of the stakes is determined, ensuring fairness regardless of the game's interruption.

PASCAL'S METHOD: USING PASCAL'S TRIANGLE

In contrast, Pascal's method takes a more direct approach that avoids the need to consider potential game progressions beyond a decisive point. It uses Pascal's Triangle, a simple number triangle in which each number is the sum of the adjacent numbers immediately above it. This provides a solution for the correct division of stakes without the need to extend the game forward.

COMPARISON OF FERMAT'S AND PASCAL'S METHODS

When applied to the same scenarios, both Fermat's and Pascal's methods yield the same results. Whether it's a scenario where Player 1 leads 2-1, 3-2, or 3-1 in a first-to-four contest, both methods correctly determine the division of the stakes in the interrupted game. Despite their different approaches, Fermat's and Pascal's solutions always align, emphasising their validity and mathematical consistency.

CONCLUSION: A MILESTONE IN PROBABILITY THEORY

The Pascal–Fermat Problem of Points signifies a turning point in the history of probability theory. By shifting focus from the history of the game to its potential outcomes, Pascal and Fermat introduced the notion of expected value, laying the groundwork for modern probability theory. Their solution also embodies the principles of fairness and proportionality, providing a theoretical framework for understanding interrupted games in a variety of contexts.

When Should We Double Up? Exploring the Martingale Betting Strategy

THE MARTINGALE BETTING STRATEGY

The Martingale betting strategy is based on the principle of chasing losses through progressive increase in bet size. To illustrate this strategy, let's consider an example: A gambler starts with a £2 bet on Heads, with an even money payout. If the coin lands Heads, the gambler wins £2, and if it lands Tails, they lose £2.

In the event of a loss, the Martingale strategy dictates that the next bet should be doubled (£4). The objective is to recover the previous losses and achieve a net profit equal to the initial stake (£2). This doubling process continues until a win is obtained. For instance, if Tails appears again, resulting in a cumulative loss of £6, the next bet would be £8. If a subsequent Heads occurs, the gambler would win £8, and after subtracting the previous losses (£6), they would be left with a net profit of £2. This pattern can be extended to any number of bets, with the net profit always equal to the initial stake (£2) whenever a win occurs.

CHASING LOSSES AND THE LIMITATIONS

While the Martingale strategy may appear promising in theory, it is important to recognise its limitations and the inherent risks involved. The strategy involves chasing losses in the hope of recovering them and generating a profit. However, it's crucial to understand that the expected value of the strategy remains zero or even negative.

The main reason behind this lies in the presence of a small probability of incurring a significant loss. In a game with a house edge, such as in a casino, the odds contain an edge against the player. The house edge ensures that, over time, the expected value of the bets is negative. Therefore, even with the Martingale strategy, which aims to recover losses, the expected value of the bets remains unfavourable.

Moreover, in a casino setting, there are structural limitations that impede the effectiveness of the Martingale strategy. Most casinos impose limits on bet size. These limits prevent gamblers from doubling their bets indefinitely, even if they have boundless resources and time, thereby constraining the strategy's potential for recovery.

THE DEVIL'S SHOOTING ROOM PARADOX

A parallel thought experiment known as the Devil's Shooting Room Paradox adds an intriguing twist. In this scenario, a group of people enters a room where the Devil threatens to shoot everyone if he rolls a double-six. The Devil further states that over 90% of those who enter the room will be shot. Paradoxically, both statements can be true. Although the chance of any particular group being shot is only 1 in 36, the size of each subsequent group in this thought experiment is over ten times larger than the previous one. Thus, when considering the cumulative probability of being shot across multiple groups, it surpasses 90%.

Essentially, the Devil's ability to continually usher in larger groups, each with a small probability of being shot, ultimately results in the majority of all the people entering the room being shot.

A key assumption underlying the Devil's Shooting Room Paradox is the existence of an infinite supply of people. This assumption aligns with the concept of infinite wealth and resources often associated with Martingale-related paradoxes. Without a boundless supply of individuals to fill the room, the cumulative probability of over 90% cannot be definitively achieved.

The Devil's Shooting Room Paradox serves in this way as another illustration of how probabilities and cumulative effects can lead to counterintuitive outcomes.

CONCLUSION: THE LIMITS OF A MARTINGALE STRATEGY

The Martingale strategy is based on chasing losses, but its expected value remains zero or negative due to the house edge. The strategy's viability is further diminished by limitations on bet size in real-world casino scenarios. As such, the Martingale system cannot be considered a winning strategy in practical gambling situations. The Devil's Shooting Room Paradox further demonstrates the complexities and counterintuitive outcomes that can arise when infinite numbers are assumed. Ultimately, a comprehensive understanding of these paradoxes provides valuable insights into the rationality of betting strategies and decision-making in the realm of gambling.

When Should We Back the Favourite? Exploring the Favourite-Longshot Bias

INTRODUCING THE FAVOURITE-LONGSHOT BIAS

The favourite-longshot bias is a persistent anomaly found in various betting markets around the globe. The bias describes a consistent pattern where bettors overvalue 'longshots', low-probability outcomes with high potential payouts, and relatively undervalue 'favourites', high-probability outcomes with low potential payouts. Put another way, bettors tend to bet too much on longshots despite their lower probability of success, and relatively too little on favourites despite their higher probability of success.

ILLUSTRATING THE FAVOURITE-LONGSHOT BIAS

The favourite-longshot bias can be better appreciated through real-world examples. To illustrate, consider the hypothetical scenario of Mr. Baker and Mr. Carpenter. Each starts with a betting bank of £1,000 and decides to place a series of £10 bets on horses with differing odds.

Mr. Baker bets on favourites. He places his flat £10 bets on 100 horses each quoted at 2 to 1 odds. This means for each £10 Mr. Baker stakes, he would gain a net £20 if his selected horse wins.

On the other hand, Mr. Carpenter, who prefers high-risk, high-return outcomes, places the same £10 bets but on 100 horses each quoted at 20 to 1 odds. This means that for every £10 bet that Mr. Carpenter places, he earns a net £200 profit if the horse wins.

Based on decades of data, we would in these circumstances expect to see Mr. Baker come out well ahead of Mr. Carpenter. Betting on favourites may not earn you a profit, but over time the data suggests that you will tend to lose a lot less than by betting an equivalent amount on longshots.

Even so, the favourite-longshot bias describes a tendency, not an ironclad rule. Although favourites are generally relatively undervalued and longshots are overvalued, some long-shots might offer very good value and some favourites might offer dreadful value.

THE HENERY HYPOTHESIS

The Henery Hypothesis, put forward by mathematician Robert Henery in 1985, offers an intriguing perspective on why the favourite-longshot bias occurs in betting markets. According to Henery, bettors have a natural tendency to discount a fixed fraction of their losses, assigning them less importance than their gains. This cognitive bias leads bettors to give less weight to losses than they should.

For instance, consider a bettor who loses a wager on a longshot horse. Even though the loss is real, the bettor might rationalise it as a 'near win' or as an outcome that was almost a victory, thereby psychologically reducing the impact of the loss. This ability to discount losses can lead bettors to prefer longshot bets because the perceived potential gain from a successful longshot bet is highly attractive, and the loss, if it occurs, can be mentally downplayed.

Importantly, the Henery Hypothesis provides a plausible explanation for the existence of the favourite-longshot bias across many different types of betting markets. If bettors universally discount a certain fraction of their losses, this would lead to a systematic overvaluation of longshots and relative undervaluation of favourites, which is the characteristic pattern of the favourite-longshot bias.

PERSISTENCE OF BIAS

Even with overwhelming evidence indicating that a consistent, level-stakes betting strategy at short odds is likely to yield higher returns than betting at long odds, the favourite-longshot bias endures. This persistence defies the expectation that market forces of supply and demand would correct this anomaly, and instead suggests that the bias is deeply ingrained in the betting markets.

There are several other theories attempting to explain the tenacity of this bias. One hypothesis is that bettors are inherently risk-loving, finding thrill and enjoyment in the high risk associated with betting on longshots. This behaviour diverges significantly from traditional financial behaviour theories, which typically assume individuals are risk-averse and prefer to minimise volatility.

An alternative explanation is that a significant number of bettors in the market may lack the necessary skills to accurately estimate the probabilities of winning associated with different odds. This could lead to a tendency to over-bet at long odds due to a misguided belief in the potential of large payoffs.

It's also possible that social factors play a role. The allure of a big win, even if it's unlikely, can be hard to resist, and social reinforcement—such as the stories of others who've won big on longshots—can contribute to the persistence of the favourite-longshot bias. Furthermore, in a culture where longshot bets are seen as bold or daring, making such bets could confer a certain social status or reputation, further encouraging the bias.

THE ROLE OF ODDS-SETTERS IN THE BIAS

Another plausible explanation for the persistent favourite-longshot bias lies in the behaviour and strategic approaches of those who set the odds—the odds-setters or bookmakers. They play a crucial role in shaping the betting market, and their influence can significantly impact the way bettors allocate their stakes.

One hypothesis is that odds-setters could be artificially manipulating odds, particularly at the longer end of the market, as a defensive mechanism against well-informed bettors. In this scenario, by reducing the odds for longshot events, bookmakers can protect themselves from potential large losses that could occur if a longshot bet with substantial stakes were to win.

In addition to this protective mechanism, it's also feasible that odds-setters are aware of the favourite-longshot bias among bettors and deliberately squeeze longer odds to exploit this tendency. If bettors are more inclined to place bets on longshots, then reducing the payout associated with these bets could improve the odds-setter's margin. Furthermore, changes in the odds at the longer end of the market may be less noticeable to bettors than at the shorter end, allowing these odds to be adjusted with less risk of deterring bettors.

CONCLUSION: A UNIFIED EXPLANATION?

The favourite-longshot bias, one of the most enduring and perplexing anomalies observed in sports betting markets, is a testament to the fascinating intersection of psychology, economics, and decision-making. Despite the plethora of evidence documenting the bias through the decades and across numerous settings, there is no unified explanation for the bias.

It is quite plausible that there is no unique explanation for the favourite-longshot bias, but rather a complex mosaic of contributory factors. The extent to which each element contributes to the bias may depend on the market structure and context.

What we do understand of the favourite-longshot bias, however, not only offers insights into the peculiarities of betting markets but also has broader implications for our understanding of human decision-making under uncertainty. It highlights how people perceive risk and reward, how they process information, and how preferences and biases can shape decisions and markets.

When Should We Expect Value? Exploring the Expected Value Paradox

UNDERSTANDING THE EXPECTED VALUE PARADOX

At its core, the Expected Value (EV) Paradox invites us to examine how outcomes deviate when we analyse them through the lens of a single ensemble event (a large group participating in an event once) vs. a multiple time event (a single individual participating in the event multiple times).

Take the example of a hypothetical coin-tossing game where players gain 50% of their bet if the coin lands on Heads and lose 40% if it lands on Tails. This game seems favourable for the player—the game has what is termed a positive expected value.

However, the paradox arises when the concept of time is introduced into the equation. While the game appears favourable in theory, it could lead to a net loss for an individual playing this game multiple times. As the coin is tossed more and more, the individual's wealth may diminish over time, leading to a scenario where they lose all their money, even though the theoretical gain from playing the game is positive.

THE EXPERIMENT

Let's set up an experiment involving a coin-tossing game with 100 participants, each with an initial stake of £10, to illustrate the difference. In this scenario, we're employing what's known as an ensemble perspective, where we're examining a large group participating in an event once.

Statistically, given a fair coin, we would expect roughly half of the coin tosses to land on Heads and half on Tails. Therefore, of the 100 people, we predict that around 50 people will toss Heads and 50 will toss Tails.

If the coin lands on Heads, each of the 50 players stands to gain 50% of their stake, which is £5. In total, this translates to a combined gain of £250 (50 players × £5).

On the other side, if the coin lands on Tails, each of the remaining 50 players loses 40% of their stake, which is £4. This accumulates to a total loss of £200 (50 players × £4).

Subtracting the total loss from the total gain (£250 − £200), we find a net gain of £50 over all 100 players. When we average this out over the number of players, we see an average net gain of £0.5 (50 pence) per player (£50 ÷ 100 players), or 5% of the £10 initial stake.

THE PARADOX

The Expected Value Paradox becomes evident when we shift from an ensemble perspective, involving many people playing the game once, to a time perspective, involving one person playing the game multiple times.

Let's examine a scenario where a single player engages in four rounds of the game, starting with a stake of £10. For simplicity's sake, we'll assume an equal chance of landing Heads or Tails—therefore expecting two Heads and two Tails.

When the coin lands on Heads in the first round, the player gains 50% of their stake, increasing their wealth to £15 (£10 + 50% of £10). If the coin lands on Heads again in the second round, their wealth grows to £22.50 (£15 + 50% of £15).

However, the game changes when the coin lands on Tails in the third round. The player loses 40% of their current wealth, reducing it to £13.50 (£22.50 minus 40% of £22.50). If the coin lands on Tails again in the fourth round, the player's wealth decreases further to £8.10 (£13.50 − 40% of £13.50).

Despite starting the game with a positive expected value, the player ends up with less money than they started with. Even though the probabilities haven't changed, the effects of winning and losing aren't symmetric.

Thus, the Expected Value Paradox is clear in this example. When many people play the game once (ensemble averaging), the average return is positive, aligning with the expected value. However, when a single person here plays the game multiple times (time averaging), the player loses money.

TIME AVERAGING AND
ENSEMBLE AVERAGING

In understanding the Expected Value Paradox, we are introduced to two different types of averaging: 'time averaging' and 'ensemble averaging'.

TIME AVERAGING

'Time averaging' is a concept that comes into play when we are observing a single entity or process over an extended period. In the context of our coin-tossing game, time averaging refers to tracking the wealth of a single player as they participate in multiple rounds of the game. Over time, this player's wealth fluctuates, often resulting in an overall loss despite the odds being in their favour. A severe loss (like bankruptcy) at any point can end the game for the player.

In our coin-tossing game, this would be akin to observing 100 players tossing the coin once. The overall gain camouflages the individual experiences, which can significantly vary—some players win, some lose.

ENSEMBLE AVERAGING

The ensemble average gives us a snapshot of the behaviour of many at a specific moment in time. The 'ensemble probability' refers to a large group's collective experiences over a fixed period.

TIME VS. ENSEMBLE AVERAGING

This difference between 'time probability' and 'ensemble probability' underscores that a group's average experience does not accurately predict an individual's experience over time.

Understanding the distinction between these two types of averaging is crucial when interpreting outcomes of games, experiments, or any process involving randomness and repetition over time. This differentiation becomes especially important in fields like economics and finance, where these principles can guide strategy and risk management.

Strategies that work on an ensemble basis may not be effective (or could be disastrous) when applied over time by an individual—a paradox manifested clearly in our coin-tossing game.

SURVIVORSHIP AND WEALTH TRANSFER

Survivorship and wealth transfer are key elements in understanding how wealth moves around in situations like gambling and investing. The term 'survivors' refers to those who keep playing the game through various rounds, while 'non-survivors' are the ones who quit, or are pushed out, often because they've lost most or all of their money.

The idea is that the wealth lost by non-survivors doesn't disappear. Instead, it gets transferred to the survivors, redistributing wealth within the system. Take a coin-tossing game as an example: if half of the 100 players lose everything and leave, while the other half double their initial amount, the group seems to break even. But, half of the players have nothing, while the other half have doubled their money.

CONCLUSION: THE INDIVIDUAL AND THE GROUP

In the conventional, or ensemble, view of probability, we look at the outcomes of many trials of an event and calculate averages. Some will win, some will lose, but overall the average outcome should reflect the true odds of the game. The individual variations or 'paths' of each person aren't considered—we're only interested in the average outcome. This so-called ensemble perspective is often used in classical statistics and probability theory. In contrast, the path-dependent view recognises that the order of events matters.

Take a person who plays a game 100 times. Even if the odds of each game are in their favour, they could still lose all their money if they have a run of bad luck. In this case, looking at the overall or ensemble average wouldn't accurately reflect the individual's experience.

In summary, while the ensemble view can provide a broad understanding of expected outcomes, the path-dependent view provides a more nuanced understanding of individual experiences.

A Question of Reason

5

When Should We Doubt Our Senses? Exploring the Probability of a Simulated Reality

---

THE SIMULATION HYPOTHESIS

Is our reality a simulation crafted by a more advanced civilisation? This provocative question, central to the Simulation Hypothesis popularised by philosopher Nick Bostrom, challenges our understanding and perceptions of existence and reality. Bostrom's concept of "ancestor simulations" proposes that a sufficiently advanced civilisation could simulate consciousness, allowing simulated beings to experience life as we know it. If such civilisations exist and choose to run these simulations, the likelihood that we're in one rather than 'base reality' increases significantly.

The creators could be located at any stage in the universe's timeline, even billions of years into the future.

It's a sort of digital time travel, allowing them to witness and potentially even interact with their own past. Bostrom argues that if any civilisation reaches a high enough technological level to be able to run these sorts of simulations and is interested in doing so, we are more likely to be in one of these simulations rather than in 'base reality'.

DOI: 10.1201/9781003402862-5

THE FOUNDATION OF BOSTROM'S SIMULATION ARGUMENT

Bostrom argues that at least one of three possibilities must be true:

1. Civilisations at our level of development almost invariably fail to reach a technologically super-advanced stage. This failure is marked by their extinction or incapacity to develop the technological means necessary to create highly detailed simulations of reality including simulated minds.
2. Among civilisations that do reach a super-advanced stage, possessing the ability to create highly detailed simulations of their ancestors or historical periods, there is an overwhelming lack of interest in actually conducting such simulations.
3. We are almost certainly existing within a simulation ourselves. This follows from the assumption that if super-advanced civilisations have both the interest and capability to run numerous simulations, the number of simulated consciousnesses would vastly outnumber the number of "real" consciousnesses.

Bostrom's argument invites us to consider the implications of our technological trajectory and the nature of consciousness. It proposes a framework where the advancement towards and the potential capabilities of super-advanced civilisations lead to a significant probability that our perceived reality might not be the base reality. This philosophical inquiry not only challenges our understanding of existence but also highlights the profound implications of future technological capabilities on our perception of reality and consciousness.

PROBING THE DEPTHS OF THE SIMULATION ARGUMENT

To fully grapple with these propositions, we must examine each statement individually. For the first proposition to be false, a civilisation would need to exhibit the capability to survive potentially catastrophic phases, whether they are caused intentionally, accidentally, or through ignorance, without succumbing to complete annihilation.

The second proposition is highly dependent on factors we can hardly predict, such as the ethical frameworks of advanced civilisations, their curiosity, and their respect for the integrity of intelligent consciousness. Even so, it might seem implausible that almost no civilisations with the capacity to create such simulations would choose to do so.

Unless civilisations either fail to reach the stage at which they can create such simulations or choose not to do so, then we must face a startling conclusion: we are very probably living in a simulation.

NAVIGATING THE PROBABILITY LANDSCAPE

Summarising the argument, a 'technologically mature' civilisation would have the capability to create simulated minds. Hence, one of the following must hold:

1. The fraction of civilisations reaching 'technological maturity' is close to zero or zero.
2. The fraction of these advanced civilisations willing to run these simulations is close to zero or zero.
3. We are almost sure to be living in a simulation.

If the first proposition holds true, our civilisation will almost certainly not reach 'technological maturity', which introduces a sense of urgency and uncertainty regarding our collective future. If the second proposition is true, then almost no advanced civilisations are interested in creating simulations, raising questions about the nature and motivations of advanced civilisations. If the third proposition is true, then we should challenge our entire perception of reality.

In the face of such profound uncertainty, we might find it pragmatic to assign equal weight to each of these propositions.

THE SINGULAR CIVILISATION HYPOTHESIS

But what if ours is the only civilisation that will ever reach our stage of development? This concept fundamentally changes the dynamics of the simulation argument. In correspondence with me, Professor Bostrom sheds some light on this question:

'If we are the only civilization at our stage there will ever have been, then the equation remains true, although some of the possible implications become less striking … the probability that we are not in a simulation is increased if ours is the only civilization that will have ever existed throughout the multiverse' (Nick Bostrom e-mail, 10 February 2021).

This assertion reinforces the complexity of the simulation argument and the profound effect that our assumptions about the universe have on our interpretations of existence.

CONCLUSION: THE PARADOX OF CREATION

The Simulation Argument presents a curious paradox. The closer we get to the point of being capable of creating our own simulations, the greater the probability that we are living in a simulation ourselves. As we stand on the precipice of creating our virtual realities, we would be faced with the startling possibility that we are simulated beings about to create a simulation.

By abstaining from creating these simulations, we could perhaps decrease the likelihood of us being simulated, indicating that at least one civilisation capable of such feats decided against it. But the moment we dive into creating simulated realities, we would be compelled to accept that we are almost certainly doing so as simulations ourselves.

This paradox inevitably leads to the obvious question: Who created the first simulation? Might that really be us? Such questions punctuate our exploration of the possibility of a simulated reality. The answers may reshape our very understanding of existence itself. But would it ultimately change anything?

When Should We Accept Unlikely Evidence? Exploring the Raven Paradox

UNDERSTANDING HEMPEL'S PARADOX

In the mid-20th century, philosopher Carl Gustav Hempel introduced a paradox that came to be known as 'Hempel's Paradox' or the 'Raven Paradox'. The paradox begins with a seemingly simple and clear premise: If the hypothesis is that 'all ravens are black', then any observation of a black raven should help to support the hypothesis.

However, Hempel pointed out that this statement is logically equivalent to the statement: 'Everything that is not black is not a raven'. Hence, any observation of a non-black, non-raven object, such as a white tennis shoe, should also help to support the hypothesis.

Yet it feels strange to believe that seeing a white tennis shoe should serve to increase our belief that all ravens are black.

HEMPEL'S PARADOX AND THE COLOUR OF FLAMINGOS

Now, let's apply this principle to another statement: 'All flamingos are pink'. This proposition is logically equivalent to: 'Everything that is not pink is not a flamingo'. By Hempel's argument, observing an object that is not pink and not a flamingo, such as a white tennis shoe, would provide evidence in support of the hypothesis that all flamingos are pink.

From a formal logic perspective, this argument makes sense. However, our intuition may still find this hard to accept, mirroring the original conflict inherent in Hempel's Paradox.

TESTING THE HYPOTHESIS

In conventional hypothesis testing, we would go out and find some flamingos, verifying if they are indeed pink. But the Raven Paradox suggests that we could conduct meaningful research by simply looking at random non-pink things and checking if they are flamingos. As we collect data, we increasingly lend support to the hypothesis that all non-pink things are non-flamingos, equivalently that all flamingos are pink.

While this approach holds up logically, it does have its limitations. Considering the vast number of non-pink things in the world compared to the population of flamingos, the hypothesis can be much more confidently validated by sampling flamingos directly. Hence, although Hempel's Paradox does not contain a logical flaw, it is not an efficient or practical method for testing the hypothesis.

THE ACCESSIBILITY PRINCIPLE (OR OBSERVATIONAL LIKELIHOOD PRINCIPLE)

Suppose we have two hypothetical species—one is a type of bird that frequents populated areas (Species A), and the other is a rare kind of lizard that lives in remote, inaccessible areas (Species B). If both these species are unobserved, it's more likely that Species B exists rather than Species A, because Species B is less likely to be observed due to its habitat even if it exists. In contrast, Species A should have been observed if it were real due to its frequent presence in populated areas. I term this the 'Accessibility Principle', or alternaively the 'Observational Likelihood Principle'. These terms suggest that the likelihood of an entity's existence depends on its observability. This aligns with real-world scientific practices, where the absence of evidence is not always evidence of absence, particularly when dealing with hard-to-observe phenomena.

So, let's take the propositions in the thought experiment in turn. Proposition 1: All flamingos are pink. Proposition 2 (logically equivalent to Proposition 1): Everything that is not pink is not a flamingo. Proposition 3 (the 'Accessibility Principle'): If something might or might not exist but is difficult to observe, it is more likely to exist than something which can be easily observed but is not observed.

Following from these propositions, when I see two white tennis shoes, I am ever so slightly more confident that all flamingos are pink than before. This is especially so if any non-pink flamingos that might be out there would be easy to spot. And I'd still be wrong, but for all the right reasons.

CONCLUSION: THE OBSERVATION PARADOX

In summary, Hempel's Paradox is an intriguing clash between intuitive reasoning and formal logic. It forces us to confront the subtleties of hypothesis testing and belief formation. In this example, the paradox implies that we may gain a tiny bit more confidence in the hypothesis that all flamingos are pink if we observe a white tennis shoe. However, such indirect evidence should be considered in its appropriate context, not as a substitute for direct evidence. The key point of the paradox is instead to challenge our understanding of the meaning of evidence and to provide valuable insights into the nature of logical reasoning. Essentially, any hypothesis is always susceptible to new evidence that can strengthen support for it. In the case of the pink flamingo hypothesis, this applies whether it comes from observing a flock of pink flamingos or (to a much lesser degree) a pair of white tennis shoes. Until you see an orange flamingo, then you know otherwise!

When Should We Accept the Simplest Explanation? Exploring Occam's Razor

THE PRINCIPLE OF SIMPLICITY

In this section we explore Occam's Razor. William of Occam (also spelled William of Ockham) was a prominent 14th-century philosopher and theologian known for his emphasis on simplicity in philosophical and theological matters. His philosophical contributions, particularly the principle of simplicity, have had a lasting impact on various fields of knowledge. Occam's Razor, derived from his philosophy, has become synonymous with the method of eliminating unnecessary hypotheses and choosing the simplest explanation consistent with the evidence.

OCCAM'S RAZOR: PRINCIPLE AND EXPLANATION

At the heart of Occam's philosophy, therefore, is the principle of simplicity, which later became known as Occam's Razor. The razor can be summarised as follows: 'Entities should not be multiplied without necessity'. In other words, when faced with competing explanations or hypotheses, the simplest one that adequately explains the available evidence should be preferred.

Occam's Razor guides our thinking by encouraging us to avoid unnecessary assumptions and complexities. It suggests that we should prefer explanations that require fewer additional elements or entities. By choosing simplicity over complexity, Occam's Razor helps us navigate knowledge acquisition and hypothesis formation.

To be clear, it's important to note that simplicity does not mean 'easier to understand' but rather 'involving fewer assumptions or conjectures'. Complexity should only be considered 'when simplicity fails to adequately explain the phenomenon.

OCCAM'S RAZOR: A CRUCIAL HEURISTIC

Occam's Razor serves as a crucial heuristic in problem-solving and theory formulation. It proposes that among competing hypotheses, the one with the fewest assumptions should be selected, provided it adequately explains the phenomenon in question.

Occam's Razor does not just simplify our thinking processes; it actively steers us away from the allure of unnecessary complexities and conjectures. By advocating for simplicity, it aids in refining our approach to knowledge acquisition and hypothesis development, ensuring that complexity is introduced only when absolutely necessary to explain the data adequately.

THE ROLE OF OCCAM'S RAZOR IN SCIENCE: TOWARDS ELEGANT EXPLANATIONS

The implications of Occam's Razor extend significantly into scientific inquiry. It underpins the scientific method, where explanations for observed phenomena are sought and hypotheses are developed. By favouring parsimonious explanations, the principle encourages scientists to construct theories that are not only elegant but also more comprehensible and testable. This preference for simplicity has facilitated remarkable advancements in our understanding of the world, emphasising that the most profound explanations often emerge from the most straightforward assumptions.

OCCAM'S RAZOR AND OVERFITTING: COMPLEXITY AND GENERALISATION

Occam's Razor finds empirical support in the phenomenon of overfitting, particularly in the field of statistics and machine learning. Overfitting occurs when a model becomes overly complex and fits the noise or random variations in the data instead of capturing the true underlying patterns.

By adhering to Occam's Razor, researchers can avoid the pitfall of overfitting, ensuring that their models capture the essential features of the data while remaining parsimonious.

OCCAM'S LEPRECHAUN: AVOIDING AD HOC HYPOTHESES

In the pursuit of explanations, it is common to encounter situations where additional assumptions are introduced to save a theory from being falsified. These ad hoc hypotheses act as patches to compensate for anomalies that were not anticipated by the original theory. Occam's Razor plays an essential role in evaluating such situations.

Imagine a situation, for example, where someone claims that a mischievous leprechaun is responsible for breaking a vase. There is likely to be serious scepticism about this claim. However, the person who made the claim introduces a series of ad hoc explanations to counter potential falsification.

For instance, when a visitor to the scene sees no leprechaun, the claimant asserts that the leprechaun is invisible. To test this, the visitor suggests spreading flour on the ground to detect footprints. In response, the claimant states that the leprechaun can float, thus leaving no footprints. The visitor then proposes asking the leprechaun to speak, but the claimant asserts that the leprechaun has no voice. In this way, the claimant keeps introducing additional explanations to prevent the hypothesis of the leprechaun's existence from being falsified.

The example of Occam's Leprechaun illustrates how additional assumptions can be added in an ad hoc manner to preserve a theory from being disproven. These 'saving hypotheses' create a flow of additional explanations that make the theory less able to be falsified. Occam's Razor encourages us to be sceptical of such ad hoc hypotheses and instead favours simpler explanations that adequately account for the evidence.

OCCAM'S RAZOR AND PREDICTIVE POWER: PARSIMONY AND EFFICIENCY

Another aspect of Occam's Razor is its association with the predictive power of theories. A theory that can accurately predict future events or observations based on fewer assumptions is considered more efficient.

By favouring simplicity, Occam's Razor guides scientists to develop theories that provide explanatory power and predictive efficiency. The simplest theory that adequately explains the available data and accurately predicts future outcomes is often preferred.

This emphasis on predictive power aligns with the pragmatic approach that Occam advocated, where theories are judged not only by their ability to explain past observations but also by their ability to make successful predictions.

CONCLUSION: THE PURSUIT OF KNOWLEDGE

In sum, Occam's Razor, a principle deeply rooted in the philosophy of William of Occam, remains a fundamental tool in the pursuit of knowledge. It encourages simplicity, efficiency, and parsimony in our explanations and theories. By guiding us towards the simplest explanations that remain consistent with the available evidence, Occam's Razor plays a crucial role in scientific methodology, theory development, and everyday reasoning. Its continued relevance underscores the timeless appeal of simplicity in our quest to understand and explain the world around us.

When Should We Believe Two Contradictory Things? Exploring the Quantum World

THE QUANTUM WORLD

Quantum physics continues to confront our vision of how the universe works and offers a mind-bending challenge to common sense. An essential part of this exploration lies in the imaginative domain of thought experiments, hypothetical scenarios that push the boundaries of our understanding. Among them, the Schrödinger's Cat and

Quantum Suicide thought experiments stand out as fascinating illustrations of quantum mechanics, each providing unique insights into the seemingly contradictory nature of the quantum world.

THE ENIGMA OF SCHRÖDINGER'S CAT

In the domain of quantum mechanics, we meet questions that upend our basic understanding of reality. One such question, famously posed by physicist Erwin Schrödinger, asks: 'Can something exist in two diametrically opposite states at the same time—such as being both alive and dead simultaneously?'

This is the core inquiry of Schrödinger's classic thought experiment.

Schrödinger's scenario involves a cat placed in an opaque box alongside a Geiger counter to detect radioactive decay, a radioactive substance, and a vial of poison. If the radioactive material decays, it sets off the Geiger counter, which in turn releases the poison and kills the cat. However, until the box is opened, it is not known whether the radioactive material has decayed. Hence, before the box is opened, one could argue that the cat is simultaneously alive, dead, or possibly both.

According to classical physics and our everyday intuition, the cat is definitively either alive or dead. Quantum mechanics proposes an entirely different narrative. In the quantum world, the cat exists in a superimposed state, being both alive and dead until someone opens the box and observes it. At this point, the superposition collapses, and the cat is forced into a single state of being either alive or dead. This symbolises the quantum state of particles existing in multiple states simultaneously, and it is a powerful metaphor for the counterintuitive nature of quantum phenomena.

PHOTONS AND THE CONUNDRUMS OF QUANTUM MECHANICS

This dichotomy between classical intuition and quantum mechanics intensifies when considering entities at the quantum level. Photons, the smallest measure of light, can simultaneously exist as both waves and particles. These states coexist until the photon is observed, at which point the photon is forced to 'decide' to be either a wave or a particle. This counterintuitive behaviour has profound implications for the way that we perceive reality.

THE COPENHAGEN INTERPRETATION AND ITS UNRESOLVED QUESTIONS

The traditional way to make sense of these quantum enigmas is through the lens of the Copenhagen Interpretation. This interpretation proposes that quantum particles exist in multiple states simultaneously, a condition referred to as superposition. Only upon observation does this superposition collapse into a single state.

But the Copenhagen Interpretation leaves us with questions as well as answers. If a quantum particle exists in multiple states at once prior to observation, what happens to all the other states that it does not collapse into upon observation? Where do these 'unobserved' states vanish to?

QUANTUM SUICIDE AND THE MANY-WORLDS INTERPRETATION

These perplexing questions pave the way for another thought experiment, known as Quantum Suicide. In this scenario, Schrödinger's cat is replaced with a man and a gun. The firing of the gun, and thus the man's survival, is determined by the spin of a quantum particle.

According to the Copenhagen Interpretation, there is a certain probability that the particle will spin one way or the other, and eventually, the man will meet his demise. However, an alternative view—the Many-Worlds Interpretation (MWI)—proposes that both outcomes are equally real and occur in separate realities.

The MWI asserts that every quantum event results in the universe splitting into different realities. Each reality embodies a possible outcome of the quantum event. In the Quantum Suicide experiment, there exists one reality in which the man dies and another reality in which he lives. This interpretation, though it goes beyond the bounds of our current understanding, identifies potential new frontiers in our quest to understand the fundamental nature of reality.

THE DEBATE OVER INTERPRETATIONS

The Copenhagen and Many-Worlds Interpretations offer drastically disparate views of the universe's fabric. The Copenhagen Interpretation embraces a unique, yet uncertain, reality that becomes concrete only when observed. Conversely, the MWI supports the idea of countless realities, each manifesting a different outcome of quantum events. These starkly different interpretations illuminate the complexity and enigma of quantum mechanics.

The Burden of Proof: Many-Worlds Interpretation

Advocates of the MWI argue that the existence of multiple, branching worlds, each harbouring a different reality, is inherently implied in the mathematical architecture of quantum mechanics.

In the realm of quantum mechanics, Schrödinger's equation governs the evolution of quantum states over time. Supporters of the MWI suggest that this equation, when followed faithfully, leads naturally to the conclusion of multiple worlds. They argue that these alternative realities exist in the quantum superposition until a measurement is made, causing the world to 'split' into distinct realities for each possible outcome.

The proponents of this interpretation consider their viewpoint as a natural, 'no-frills' approach to quantum mechanics. They contend that it's merely accepting what the mathematical formalism of quantum theory is suggesting—an unending tree of branching universes, each representing a possible outcome of a quantum event.

The Burden of Proof: Copenhagen Interpretation

On the other hand, supporters of the Copenhagen Interpretation propose that evidence should be grounded in the observable world. They contend that the onus of proving the existence of alternate realities lies on the advocates of the MWI. The Copenhagen school considers the quantum wave function, which is at the heart of the quantum mechanics formalism, as providing a probabilistic description of reality rather than a deterministic one.

In this perspective, quantum superposition allows for a range of potentialities, but only one is realised upon observation. This interpretation argues that there is only one world—the world we can observe and measure. The wave function is a mathematical tool that enables the calculation of probabilities for different outcomes but doesn't necessitate the existence of alternate realities.

CONCLUSION: QUANTUM PARADOXES

Where we land in this debate might be significantly influenced by our choice of default stance. Are we ready to entertain the idea of countless, concurrent realities, a notion seemingly out of a science fiction novel? Or do we feel more grounded in the tangible, observable world, even if it's imbued with inherent uncertainty until measured?

The MWI, while appearing counterintuitive and borderline fantastical, could be the most rational explanation in some alternate reality. Conversely, the Copenhagen Interpretation, despite its clash with classical determinism, might resonate with our everyday experience of a singular, observed reality.

Ironically, it's the same quantum mechanics that allows for these starkly different interpretations to coexist—not unlike Schrödinger's cat, both alive and dead at the same time—in a superposition of states. It encapsulates the sheer wonder and paradoxes of quantum theory that continue to challenge and expand our understanding of the universe.

When Should We Think We're Special?
Exploring the Cosmos and Our Place in It

THE PRECISION OF THE COSMOLOGICAL CONSTANT

The cosmological constant, which can be interpreted as the energy density of 'empty' space, exhibits an extraordinarily precise value. If it were to differ even fractionally—being minutely larger or smaller—the universe as we comprehend it wouldn't exist.

Were the constant just a tiny bit larger, the universe would have expanded too quickly, preventing the formation of galaxies and stars. Conversely, if the constant were a shade smaller, the universe would have been crushed under the force of gravity, well before life had a chance to evolve. This delicate equilibrium of the cosmological constant underscores the argument for the universe's fine tuning.

THE INTRICATE DANCE OF FUNDAMENTAL CONSTANTS

The fundamental constants of nature are the pillars that hold up our understanding of the universe. They are the bedrock of our physical laws, governing everything from the scale of galaxies to the behaviour of subatomic particles. They also include the gravitational constant, and those associated with the strong and weak nuclear forces.

GRAVITY: THE COSMIC ARCHITECT

Gravity, quantified by the gravitational constant (G), is the force that sculpts the large-scale structure of the universe. It's the invisible hand that moulds galaxies, ignites stars, and sets planets in their orbits. This force is not as strong as the other fundamental forces, but it is long-range and dominates on the large scales of stars, galaxies, despite its relative weakness at smaller scales. As such, it has been described as the principal architect of the cosmos.

THE STRONG NUCLEAR FORCE: THE GLUE OF MATTER

The strong nuclear force is what holds the atomic nucleus together. It's an incredibly powerful force, much stronger than electromagnetism and gravity, but its influence extends only over tiny, subatomic distances. Without it, protons and neutrons wouldn't bind together in atomic nuclei, erasing the possibility of atomic, and consequently, material existence.

THE WEAK NUCLEAR FORCE: A CATALYST OF CHANGE

The weak nuclear force is responsible for processes such as radioactive decay and nuclear fusion in stars. Although it's much weaker than the strong nuclear force and electromagnetism, the weak force plays a crucial role in the nuclear reactions that power the sun, providing the energy that sustains life on Earth.

THE SENSITIVE BALANCE OF THE FUNDAMENTAL CONSTANTS

The values of these constants are finely tuned. If the strong force were a tiny bit weaker, atomic nuclei couldn't hold together, thwarting the formation of atoms. If the weak force were a tiny bit different, the sun and stars would not provide the stable energy sources necessary for life, or the potential for life, to evolve. If gravity were a tiny bit stronger, stars would burn through their fuel too quickly and violently, undermining the conditions conducive to life's evolution. This intricate balance highlights the nuanced and calibrated nature of the universe, echoing the complex interdependencies and fine-tuning essential for life and existence.

SYMMETRY AND ASYMMETRY IN THE UNIVERSE

Symmetry and asymmetry also play fundamental roles in the laws governing our Universe. Symmetry, in a broad sense, refers to any transformation that leaves a physical system unchanged. Asymmetry, on the other hand, signifies an imbalance or a deviation from perfect symmetry. The interplay between these two opposing principles shapes the Universe as we know it.

CONSEQUENCES OF PERFECT SYMMETRY AND MINIMAL ASYMMETRY

Had our Universe been perfectly symmetric, every particle of matter would have been matched with a corresponding particle of antimatter. This pair would annihilate upon contact, leaving behind nothing but radiation. Therefore, a perfectly symmetric universe would contain no matter—only energy. On the other hand, even a minimal asymmetry in the early Universe tips the scale slightly in favour of matter over antimatter. This imbalance is enough to leave behind the matter that forms stars, planets, and ultimately, us. Without this minimal asymmetry, we as well as the stars and planets would not exist.

THE PERPLEXING FINE-TUNING: A COINCIDENCE OR A PUZZLE?

TRIVIALISING THE FINE-TUNING PROBLEM: A MISGUIDED APPROACH

Some argue that the fine-tuning problem is trivial. They argue that if the Universe were any different, we simply wouldn't be here to observe it. However, this approach is unsatisfying. Let's say, for example, that I survived unharmed from a fall out of an airliner onto tarmac. That would demand an explanation. To say that I couldn't have asked the question if I hadn't survived the fall is no real argument at all. We really do need a better answer to the question of why the Universe is fine-tuned to allow for life in the first place.

A UNIVERSE OF COIN TOSSES: THE ODDS AGAINST US

To visualise the extent of the Universe's fine-tuning, we can use the analogy of a coin toss. Imagine tossing a coin hundreds or thousands of times and having it land heads followed by tails every single time, or in any other pre-determined sequence. The odds of this happening by chance are virtually zero. The same general argument applies to the odds that all the Universe's fundamental constants would align to create a life-supporting cosmos. Or say that an alien species is devising a lottery draw with a thousand balls, and the only way that humans can avoid extinction is for the balls to come out in order from 1 to 1,000. Now, that's possible, as any sequence is as likely as any other, but it's all but impossible to have happened by chance. A much more realistic and sensible conclusion is that the aliens had rigged the draw to allow us to survive.

THE JACK AND JILL CHALLENGE AND OTHER ANALOGIES

The 'Jack and Jill Challenge' is another popular analogy to highlight the fine-tuning conundrum. Imagine a Universe (let's call it Jack) that can only create life if certain parameters are precisely set. Now imagine another Universe (let's call it Jill) where life arises easily, regardless of the values of those parameters. If you had to bet on which Universe we live in, you'd most likely pick Jill, because it doesn't require any improbable fine-tuning. But we live in a Universe much more like Jack, which makes our existence seemingly miraculous.

The fine-tuning problem, the balance of matter and antimatter, and the precision of the cosmological and fundamental constants present us, therefore, with a profound puzzle. They compel us to question our understanding of the Universe and push the boundaries of our knowledge about the cosmos.

LOOKING BEYOND A SINGLE UNIVERSE: THE MULTIVERSE THEORY

In an attempt to make sense of the universe's fine-tuning, some physicists have proposed the multiverse theory. This theory proposes that our universe might not be the only one, but rather one of multiple potential universes, each with different initial conditions and possibly different laws of physics. According to this theory, each universe could have had its own Big Bang, leading to a range of universes with varying characteristics.

ADDRESSING POTENTIAL ISSUES WITH THE MULTIVERSE THEORY

While the multiverse theory is interesting, it is not without its issues. For one, it is largely speculative and not currently empirically verifiable. Critics argue that without observational or experimental evidence, the multiverse theory is more of a philosophical idea than a scientific one. Furthermore, the existence of other universes simply shifts the fine-tuning issue, questioning why the multiverse, if it exists, allows for the existence of habitable universes. Put another way, if we live in a multiverse with a landscape of varying laws, how did this landscape come to be? What mechanism generates the vast array of universes with such diversity? And why does this mechanism allow for such a distribution of universes that at least one (that we know of) can support life?

These questions touch on deep, unresolved issues at the intersection of physics, cosmology, and philosophy, highlighting the rich complexity and profound mystery of our universe.

EXAMINING THE IMPROBABILITY OF LIFE'S EMERGENCE AND THE CONCEPT OF THE GREAT FILTER

The existence of life, let alone intelligent life, in the universe adds another layer of complexity to the fine-tuning problem. Not only must the universe have the right physical laws and constants, but a series of highly improbable events must also occur for life to emerge.

TRACING BACK TO LUCA
(LAST UNIVERSAL COMMON ANCESTOR)

When considering life on Earth, we can trace all living things today back to a simple single-cell progenitor, known as the Last Universal Common Ancestor (LUCA). It's widely accepted that this extraordinary transition has occurred only once on Earth. This microscopic life-form, thought to have existed between three and four billion years ago, is a testament to the sheer improbability and remarkable evolution of life from simplicity to the complexity we witness today. Without LUCA, there would be no life as we know it, if at all. There is also the question of the origin of life itself. The transition from non-living to living matter (abiogenesis) is a major unresolved question in science.

THE GREAT FILTER

The term 'Great Filter' denotes the sequential hurdles from simple, lifeless matter to a complex, observable civilisation, of how life, in any form, could emerge from non-living material and evolve into human consciousness. The underlying theory suggests that nearly all initial attempts to progress along this trajectory encounter insurmountable obstacles— either through one exceptionally difficult stage or numerous incredibly challenging ones.

CONCLUSION: THE BIG QUESTIONS

While our understanding of the universe and its origins has advanced tremendously, the fine-tuning problem reminds us of how much we still don't know. The sheer complexity and precision of the universe's properties continue to fascinate physicists, biologists, philosophers, and theologians alike. The enigma of the universe's fine-tuning encourages us to further explore and question the nature of reality. As we continue to unravel the mysteries of the cosmos, we must keep asking: How and why does life exist? How and why did consciousness emerge from non-consciousness? Why was there originally something rather than nothing? Why is there now something rather than nothing, life rather than the absence of life?

These are very big questions. But perhaps the most amazing thing lies not in the questions themselves, but in the fact that we can ask the questions at all.

A Question of Truth

6

When Should We Question Identity?
Exploring the Ship of Theseus Paradox

HISTORICAL AND CONCEPTUAL CONTEXT

The Ship of Theseus Paradox has its roots in ancient Greek philosophy, emerging as a crucial discussion point in understanding identity and change. Originally posed by the philosopher Plutarch, the paradox was used to question whether a ship, which was gradually having all its wooden parts replaced, remained fundamentally the same ship. This paradox was not just a mere intellectual exercise; it was deeply rooted in the Greek exploration of 'being' and 'becoming', which were crucial themes in their philosophical inquiries. Over time, the Ship of Theseus became a pivotal reference in philosophical discussions about identity, persisting through the centuries as a tool to test the limits of our understanding of continuity and change.

A QUESTION OF IDENTITY

The Ship of Theseus Paradox is central to discussions in philosophy regarding the nature of identity. It presents a compelling challenge to the idea of persistent identity over time, particularly when an object undergoes gradual change.

THE LEGEND

The story of Theseus's Ship begins with the legendary hero Theseus, who sailed on a ship to the island of Crete to defeat the Minotaur. After his victory, his ship was preserved and displayed in Athens as a symbol of the city's pride. Over time, the wooden planks of the ship began to decay and were replaced with new ones. Eventually, every original piece of the ship was replaced, leading to the question: Is the ship still the same ship that Theseus sailed on, even though none of its original components remain?

CONTINUITY AND IDENTITY

If an object has all its parts replaced, is it still the same object? If we say that it is the same object, then we must explain why and how it retains its identity despite having none of its original components. Conversely, if we say that it is not the same object, then we must determine at what point it ceased to be the original and became something new.

The question of whether an object remains the same when its parts have been entirely replaced makes us reassess our understanding of what constitutes an object's identity. Are objects defined by the matter of which they're composed, their structure, their history, or by a combination of these and maybe other factors?

THE SUBSTANCE VIEW

The Substance View proposes that the identity of an object is tied to the substance or the matter it is made of. According to this perspective, the Ship of Theseus depends on the continuity of the material components that constitute it. When all the original parts of the ship are replaced, the ship loses its original identity and becomes a new object. This view sees the ship's identity as static, fixed, and dependent on its material constituents.

This interpretation faces challenges when considering gradual transformations, as it becomes difficult to pinpoint the exact moment when the ship's identity changes. Moreover, this view might struggle to account for the importance of functional and relational aspect of objects. Critics argue that it cannot satisfactorily explain cases where an object's function and relation to the world remain constant despite material changes.

Recent debates have also brought into question the implications of digital and virtual identities. In a digital era, where replication and modification of virtual entities are commonplace, how does the Ship of Theseus Paradox inform our understanding of digital identity? Does a digital object lose its 'identity' when its code is altered or does it transcend traditional notions of materiality?

THE RELATIONAL VIEW

The Relational View focuses on the idea that the identity of an object is grounded in its relationships with other objects and entities.

Supporters of the Relational View argue that the Ship of Theseus retains its identity through its connections to the story of Theseus, its role in the society in which it exists, and the memories and associations that people have with it.

THE BUNDLE THEORY

The Bundle Theory suggests that an object is nothing more than a bundle of its properties—there's no 'object' beyond the collection of its characteristics. Applying this theory to the Ship of Theseus, one might argue that the ship is merely a bundle of its properties such as its shape, size, purpose, and the arrangement of its planks. As these properties change (when the planks are replaced), the ship's identity changes too. However, if the ship retains its structure, function, and perhaps other properties, it can still be recognised as the 'same' ship. This interpretation encourages us to think of objects as collections of properties rather than stable, unchanging entities.

ARTEFACTS

In the context of the Ship of Theseus Paradox and the discussion on identity and change, the restoration of historical artifacts offers a compelling parallel.

Restoration and Identity

The process of restoring historical artefacts often involves repairing or replacing deteriorated components with new materials to preserve the artefact's appearance, function, or structural integrity. This process raises questions similar to those in the Ship of Theseus: does an artefact maintain its original identity after restoration, especially when significant portions have been replaced or altered?

Authenticity vs. Preservation

The challenge in artefact restoration lies in balancing authenticity with preservation. Authenticity refers to the degree to which an artefact remains unchanged, retaining its original materials and form. On the other hand, preservation might require the introduction of new materials to prevent further decay or to restore an artefact to a former state. At what point does an artefact become a replica rather than an original?

CASE STUDIES

The Sistine Chapel

Consider the restoration of the Sistine Chapel ceiling, where layers of grime and soot were removed to reveal Michelangelo's original colours. Some critics argued that the vibrant colours revealed by the restoration were inconsistent with Michelangelo's intentions, suggesting that the restoration had altered the fresco's identity. Others contended that the restoration brought the artwork closer to its original state, thus preserving its true identity.

The Parthenon

Similarly, the restoration of ancient buildings, like the Parthenon in Athens, involves replacing eroded stones with new material. Critics might question whether the building maintains its original identity after such changes, while proponents argue that restoration helps preserve the structure's historical and cultural significance. A key issue is whether it is more 'genuine' as a ruin bearing the marks of its history or restored to a state believed to be true to its original form.

The Last Supper

"The Last Supper" by Leonardo da Vinci has undergone several restorations over the centuries due to deterioration caused by environmental factors, wartime damage, and previous restoration attempts. Each restoration has presented a dilemma, requiring restorers to decide whether to attempt to revert the mural to its original state (as much as possible) or to stabilise its condition to prevent further degradation.

Critics argue that each layer of restoration moves the painting further from Leonardo's original vision, potentially altering its identity. They contend that the original materials, brushstrokes, and techniques employed by Da Vinci contribute fundamentally to the painting's essence and that replacing or significantly altering these elements diminishes the work's authenticity.

In philosophical terms, the restoration of "The Last Supper" mirrors the Ship of Theseus Paradox by raising questions about continuity and identity over time. If all the original pigment is removed and replaced, is it still the same painting? Or does the essence of the artwork lie in its visual appearance, its historical significance, or the intent behind its creation?

The Bridge at Mostar

Originally built in the 16th century by the Ottomans, the Stari Most stood as a symbol of unity and an architectural marvel, connecting the diverse communities in Mostar across the Neretva River. Its wartime destruction in 1993 became a poignant symbol of cultural and communal fragmentation,

The decision to rebuild the Stari Most was fraught with questions about identity and authenticity. Could a reconstructed bridge, built centuries after the original, serve the same symbolic and functional roles as its predecessor? The reconstruction effort aimed to use original techniques and materials as much as possible, sourcing local stone and employing traditional Ottoman construction methods. This approach sought to preserve the bridge's historical authenticity and cultural significance, even as it acknowledged the impossibility of an exact physical replica.

The Stari Most's reconstruction challenges the Ship of Theseus Paradox by asking whether an object—destroyed and subsequently rebuilt with the intent of mirroring the original as closely as possible—retains its identity. This case pushes the paradox further by introducing the element of complete destruction rather than gradual replacement. Is the new bridge the same as the old, despite the interruption of its physical existence? Or does its reconstruction, imbued with the collective memory, effort, and intention to bridge past and present, confer upon it a renewed identity that is both continuous and distinct?

Through its destruction and reconstruction, the Stari Most offers a powerful narrative on the complexities of identity, continuity, and change. It exemplifies how reconstructed heritage can carry forward the essence of the original, serving as a bridge not only in physical space but in time, memory, and meaning, thereby engaging with the philosophical inquiries posed by the Ship of Theseus Paradox in a deeply human context.

VIRTUAL IDENTITIES

Recent debates have also brought into question the implications of digital and virtual identities. In a digital era, where replication and modification of virtual entities are commonplace, how does the Ship of Theseus Paradox inform our understanding of digital identity? Does a digital object lose its 'identity' when its code is altered or does it transcend traditional notions of materiality?

Digital Personas

Digital personas are curated representations of ourselves on the internet, shaped by the information we choose to share on social media, forums, and other online platforms. These personas are not static; they evolve as we update our profiles, post new content, and interact with others. This fluidity raises questions akin to those posed by the Ship of Theseus: if a digital persona is constantly changing, at what point does it become fundamentally different from its original incarnation? Moreover, the curated nature of digital personas prompts us to consider which aspects of our identity are essential and which are mutable.

Artificial Intelligence

AI presents a more complex challenge to traditional concepts of identity. Machine learning algorithms allow AI systems to evolve based on new data and experiences, much like humans learn and change over time. This adaptability leads to questions about the continuity of identity: if an AI's decision-making processes and behaviours change significantly, is it still the 'same' AI?

IMPLICATIONS FOR PERSONAL IDENTITY

In terms of personal identity, the Ship of Theseus Paradox intersects significantly with theories of psychological continuity. According to this theory, personal identity is maintained through the continuity of psychological features like memory, personality, and consciousness. If we apply this to the Ship of Theseus, it raises the question: Is identity maintained through physical continuity or through the continuity of function and recognition?

This perspective is particularly relevant in discussions about human development and change. As individuals undergo physical, emotional, and psychological changes throughout life, at what point do they become 'different' individuals, if at all? The Ship of Theseus Paradox, thus, serves as a metaphor for exploring the fluidity and resilience of personal identity amidst constant change.

By integrating these aspects, the discussion around the Ship of Theseus Paradox becomes not only more historically grounded and analytically rich but also deeply connected to contemporary and personal contexts.

BROADER IMPLICATIONS

The Ship of Theseus Paradox may provide a useful framework for grappling with emerging ethical and philosophical issues as advancements in technology push the boundaries of what is possible. For instance, questions about the continuity of consciousness and the identity of entities that undergo substantial change arise in fields such as artificial intelligence and human augmentation.

CONCLUSION: CHALLENGING OUR ASSUMPTIONS

The Ship of Theseus Paradox remains an engaging and relevant tool for philosophical inquiry. Its exploration of identity and change continues to resonate with modern audiences. By challenging our assumptions and forcing us to question our understanding of the world, the paradox fulfils a key purpose of any paradox: to provoke thought and inspire exploration.

When Should We Expect Visitors? Exploring the Fermi Paradox

WHERE IS EVERYBODY?

In the early 1950s, the world was on the cusp of the Space Age, with rapid advancements in rocketry and a growing fascination with outer space. It was a time of optimism and curiosity about the cosmos, fuelled by science fiction and the nascent space programmes. Enrico Fermi, a Nobel Prize-winning physicist known for his work on the Manhattan Project, posed a question during a casual lunch conversation with colleagues, sparking a debate that would extend far beyond that moment. Fermi's question "Where is everybody?" resonated deeply. It juxtaposed the era's technological optimism with a sobering, profound mystery. Given the vastness of the universe, why is there no evidence or contact with any extra-terrestrial civilisations?

THE AGE AND SIZE OF THE UNIVERSE

The age and size of the universe are key aspects of the Fermi Paradox. The universe is approximately 13.8 billion years old, and the Milky Way galaxy, where our solar system resides, is about 13.6 billion years old. By comparison, the Earth is about 4.5 billion years old. This vast timescale implies that if the evolution of life and development of technological civilisations is a common process, there should have been ample time for numerous advanced civilisations to arise in our galaxy alone.

The sheer size of the universe reinforces this idea. The Milky Way is home to at least 100 billion stars, many of which are likely to host their own planetary systems due to the prevalence of elements necessary for planet formation. This gives rise to a very large number of potential sites for life.

Recent astronomical discoveries have further accentuated the perplexity of the Fermi Paradox. The launch of telescopes like Kepler and TESS has led to the identification of thousands of exoplanets, many of which are in the habitable zone of their stars, and it has only deepened the enigma of the paradox, making the silence in the cosmos even more confounding.

TECHNOLOGICAL ADVANCEMENT AND SINGULARITY

The concept of technological singularity—a point where technological growth becomes uncontrollable and irreversible—presents a fascinating intersection with the Fermi Paradox. The singularity is often associated with the emergence of superintelligent AI, which is theorised to have the ability to enhance its own capabilities continuously and rapidly. If other civilisations have reached singularity, leading to exponential growth in their capabilities, why is there no evidence of their existence? This discrepancy raises questions about the nature of advanced civilisations and their technological trajectories. If we consider the rapid pace of human technological development, it's reasonable to think that an extra-terrestrial civilisation, with a head start of even a few thousand years, would have achieved technological feats beyond our comprehension.

Could it be that the very nature of singularity leads civilisations to evolve in ways that are undetectable to us, or perhaps, that the pursuit of singularity inadvertently leads to self-destruction?

PROPOSED SOLUTIONS

The Zoo Hypothesis is one proposed solution to the Fermi Paradox. It suggests that extra-terrestrial civilisations are aware of our existence but have intentionally chosen not to contact us but perhaps to observe us. This could be due to a policy of non-interference, aimed at allowing younger civilisations like ours to develop and evolve independently.

The Great Filter hypothesis proposes that there is a critical barrier or a series of barriers that drastically reduce the probability of intelligent life arising, persisting, and becoming detectable by others. The concept of the Great Filter helps explain the lack of observed extra-terrestrial civilisations by suggesting that one or more critical steps in the development of life or civilisation are extremely unlikely or have a high probability of self-destruction.

A related hypothesis is the Rare Earth Hypothesis, which suggests that while simple life forms might be relatively common, more complex, multicellular organisms are exceptionally rare.

The Transcension Hypothesis offers a different perspective on the Fermi Paradox. It proposes that advanced civilisations might not expand outwards into the cosmos but rather inwards, by miniaturising and compressing their technological and informational systems. As a civilisation advances, it might focus on developing virtual realities, advanced simulations, and artificial intelligence rather than pursuing interstellar travel and communication.

In summary, the Zoo Hypothesis, the Great Filter and Rare Earth hypotheses, and the Transcension Hypothesis represent proposed solutions to the Fermi Paradox. Each offers a distinct perspective on the current lack of observed evidence of intelligent life beyond Earth.

CONCLUSION: THE SEARCH GOES ON

A number of hypotheses have been proposed to attempt to solve the Fermi Paradox. Each offers a distinct perspective on the lack of observed evidence of civilisations or intelligent life beyond our planet. These 'solutions' explore possibilities such as intentional non-interference, the existence of insurmountable barriers in the development of complex life or civilisation, and the focus on inward technological advancement. While none of these hypotheses provide a definitive answer to the Fermi Paradox, they contribute to the ongoing discussion and encourage further exploration and research in the search for extra-terrestrial intelligence in our galaxy and beyond

In conclusion, the Fermi Paradox and its related hypotheses serve not only as scientific inquiries but also as philosophical and ethical touchstones for humanity. They encourage us to ponder our existence, our future, and our responsibilities in the cosmic arena. As we continue to explore the universe and search for answers, it is possible that we may ultimately learn more about ourselves than about the cosmos that surrounds us.

When Should We Question Infinity? Exploring an Ancient Paradox

INTRODUCTION

Born in the 5th century BC in Elea (a Greek colony in southern Italy), Zeno of Elea is one of the most intriguing figures in the field of philosophy. Zeno's paradoxes are a set of problems generally involving distance or motion. While there are many paradoxes attributed to Zeno, the most famous ones revolve around motion and are extensively discussed by Aristotle in his work, 'Physics'. These paradoxes include the Dichotomy paradox (that motion can never start), the Achilles and the Tortoise paradox (that a faster runner can never overtake a slower one), and the Arrow paradox (that an arrow in flight is always at rest). Through these paradoxes, Zeno sought to show that our common-sense understanding of motion and change was flawed and that reality was far more complex and counterintuitive.

The Achilles and the Tortoise paradox, as one example, uses a simple footrace to question our understanding of space, time, and motion. While it's clear in real life that a faster runner can surpass a slower one given enough time, Zeno uses the race to craft an argument where Achilles, no matter how fast he runs, can never pass a tortoise that has a head start. This thought experiment forms a remarkable philosophical argument that challenges our perceptions of reality and creates a fascinating paradox that continues to engage scholars to this day.

These paradoxes might seem simple, but they invite us into deep philosophical waters, questioning our perception of reality and illustrating the complexity of concepts we take for granted like motion, time, and distance. In this way, Zeno's contributions continue to have profound relevance in philosophical and scientific debates, encouraging us to critically explore the world around us.

THE PARADOX OF THE TORTOISE AND ACHILLES

In one version of this paradox, a tortoise is given a 100-metre head start in a race against the Greek hero Achilles. Despite Achilles moving faster than the tortoise, the paradox argues that Achilles can never overtake the tortoise. As Aristotle recounts it, 'In a race, the quickest runner can never overtake the slowest, since the pursuer must first reach the point whence the pursued started, so that the slower must always hold a lead'.

THE UNDERLYING INFINITE PROCESS

This paradox lies in the infinite process Zeno presents. When Achilles reaches the tortoise's original position, the tortoise has already moved a bit further. By the time Achilles reaches this new position, the tortoise has again advanced. This sequence of Achilles reaching the tortoise's previous position and the tortoise moving further seems to continue indefinitely, suggesting an infinite process without a final, finite step. Zeno argues that this eternal chasing renders Achilles incapable of ever catching the tortoise.

A MATHEMATICAL SOLUTION TO THE PARADOX

The resolution to Zeno's paradox lies in the mathematical understanding of infinite series. Using a stylised scenario where Achilles is just twice as fast as the tortoise (it's a very quick tortoise!), we define the total distance Achilles runs (S) as an infinite series: $S = 1$ (the head start of the tortoise) $+ 1/2$ (the distance the tortoise travels while Achilles covers the head start) $+ 1/4 + 1/8 + 1/16 + 1/32 \ldots$

By mathematical properties of geometric series, this infinite series sums to a finite value. In other words, despite there being infinitely many terms, their sum is finite: $S = 2$. Hence, Achilles catches the tortoise after running 200 metres, demonstrating how an infinite process can indeed have a finite conclusion.

PHILOSOPHICAL IMPLICATIONS: IS AN INFINITE PROCESS TRULY RESOLVED?

Zeno's paradoxes, while they might be resolved mathematically, open a Pandora's box of philosophical questions, particularly concerning the nature of infinity and the real-world interpretation of mathematical abstractions. How can a seemingly infinite process with no apparent final step culminate in a finite outcome?

The Thomson's Lamp thought experiment, proposed by philosopher James F. Thomson, provides an insightful analogy. Imagine you have a lamp that you can switch on and off at decreasing intervals: on after one minute, off after half a minute, on after a quarter minute, and so forth, with each interval being half the duration of the previous one. Mathematically, the total time taken for this infinite sequence of events is two minutes. However, a critical philosophical question emerges at the end of the two minutes: is the lamp in the on or off state?

This question is surprisingly complex. On the one hand, you might argue that the lamp must be in some state, either on or off. However, there is no finite time at which the final switch event takes place, given the infinite sequence of switching. Hence, the state of the lamp appears indeterminate, raising questions about the applicability of infinite processes in the physical world. More prosaically, of course, you may just have blown the bulb!

This conundrum mirrors the situation in Zeno's paradox of Achilles and the Tortoise. Just as the state of Thomson's Lamp after the two-minute mark seems ambiguous, so does the concept of Achilles catching the tortoise after an infinite number of stages. While mathematics gives us a definitive point at which Achilles overtakes the tortoise, the philosophical interpretation of reaching this point through an infinite process is not as clear-cut.

The Thomson's Lamp thought experiment highlights that while we can use mathematical tools to deal with infinities, interpreting these results in our finite and discrete physical world can be philosophically challenging. It reminds us that philosophy and mathematics, while often harmonious, can sometimes offer different perspectives on complex concepts like infinity, sparking ongoing debates that fuel both fields.

ZENO'S PARADOXES, THE QUANTUM WORLD AND RELATIVITY

Zeno's paradoxes, which have puzzled thinkers for millennia, find surprising echoes in the realms of quantum mechanics and the theory of relativity, two foundational components of modern physics. These paradoxes, originally aimed at challenging the coherence of motion and time, intersect with quantum and relativistic concepts in thought-provoking ways.

In quantum mechanics, the principle of superposition allows particles to exist in multiple states at once until observed. This phenomenon reflects the essence of Zeno's Arrow Paradox, where an arrow in flight is paradoxically motionless at any instant. This comparison highlights how quantum theory disrupts traditional views on motion, suggesting that at a microscopic level, movement doesn't conform to our standard or philosophical expectations.

Meanwhile, the theory of relativity introduces the concept of time dilation, where time appears to 'slow down' for an object moving at speeds close to the speed of light. This idea provides a modern perspective on Zeno's Dichotomy Paradox, which argues that motion is impossible due to the infinite divisibility of time and space. Through relativity, we see that motion and time are relative, not absolute, concepts—illustrating a deep connection to Zeno's philosophical challenges, even after over two millennia.

CONCLUSION: PHILOSOPHICAL DEBATE AND CONTEMPORARY RELEVANCE

Contemporary philosophers continue to grapple with Zeno's paradoxes, not only as historical curiosities but also as fundamental challenges to our understanding of reality. These paradoxes force us to reconsider how we conceptualise time, space, and motion. They remind us that our intuitive grasp of the world is often at odds with its underlying complexities. In today's world, where scientific and technological advancements continually push the boundaries of what we understand, Zeno's paradoxes remain as relevant as ever, reminding us of the enduring power and limits of human reason and the ongoing journey to comprehend the universe in which we live.

When Should We Tip the Staff?
Exploring the Bell Boy Paradox

UNDERSTANDING THE CLASSIC BELL BOY PARADOX

The paradox begins when three salesmen arrive at a hotel and agree to split the cost of a room priced at £30, resulting in each person paying £10. Later, they are informed of a £5 discount, which should be returned to them. However, the bell boy, who is entrusted with returning this £5, retains £2 for himself and only refunds £1 to each guest.

This implies that each guest paid £9, making a total of £27, and with the £2 the bell boy kept, the total is £29, thus creating a seemingly missing pound.

THE ORIGINAL TRANSACTION

In the original transaction, the guests paid a total of £30 for the room, effectively transferring the amount from the guests to the hotel. Therefore, in a strictly accounted world, the hotel would have £30 and the guests would have £0.

THE REFUND SCENARIO

Upon the realisation of a discount, the manager instructed the bell boy to return £5 to the guests. However, instead of dividing the entire £5 among the guests, the bell boy takes a detour from honesty and keeps £2 for himself. He hands over £1 to each guest, meaning that the guests now have paid £9 each (totalling £27) and the bell boy has an extra £2.

DISCREPANCY IN ACCOUNTING

The confusion, and thereby the paradox, arises when you try to add the £27 that the guests effectively paid and the £2 the bell boy kept, coming up to £29, not the initial £30 that was paid. Where is the missing pound?

RESOLVING THE PARADOX

The paradox plays on a common misdirection in accounting. Adding the £2 kept by the bell boy to the £27 paid by the guests is a misleading calculation, as the £27 already includes the £2 kept by the bell boy.

When you add the £27 that the guests paid to the £2 the bell boy kept, you're effectively counting the £2 twice. The correct calculation is to subtract the bell boy's £2 from the £27 the guests paid, which gets you back to the original £25 that the room cost. The remaining £5 is accounted for in the £1 refund each guest received (totalling £3) and the £2 the bell boy kept.

CONCLUSION: LESSONS FROM THE BELL BOY PARADOX

The Bell Boy Paradox is a classic example of how a flawed understanding of a situation can lead to misconceptions and errors. In this case, the confusion arises from incorrectly adding what should have been subtracted, which is the correct method to account for the bell boy's tip. This fun riddle provides a valuable lesson in the importance of careful accounting. Remember this next time you decide to split the bill!

When Should We Flip A Card? Exploring the Four Card Problem

THE FOUR CARD PROBLEM

The Four Card Problem, also known as the Wason selection task, is a captivating puzzle that tests our logical reasoning abilities. Invented by Peter Cathcart Wason, this task challenges us to determine the minimum number of cards required to verify or falsify a given statement. Let's look deeper into this intriguing problem.

THE SCENARIO: CARD SETUP

Imagine being presented with four cards, each displaying either a letter or a number. These cards lay the foundation for the puzzle, providing the information necessary to reach a conclusion. Let's examine an example:

The face-up sides of the cards show: 23; 28; R; B

Alongside these cards, you are given a statement: 'Every card with 28 on one side has R on the other side'.

DETERMINING THE MINIMUM NUMBER OF CARDS

Now, the crucial question arises: How many cards must you turn over to determine the truthfulness of the given statement? And which specific cards should you investigate?

COMMON MISCONCEPTIONS

At first glance, the task might appear deceptively simple. Many individuals are inclined to turn over the R card, assuming it holds the key to verifying the statement. However, this line of thinking is misguided. Regardless of what is on the other side of the R card, it does not contribute to determining whether every card with 28 on one side has R on the other.

Similarly, the inclination to turn over the 23 card is also misleading. Even if the 23 card reveals an R on its other side, it does not provide any insight into the truthfulness of the statement. The existence of R on the opposite side of the 23 card merely confirms that the statement 'Every card with 23 on one side has B on the other side' is false. It does not shed light on the validity of the statement regarding the 28 card and R.

THE KEY TO SOLVING THE PUZZLE: LOGICAL ANALYSIS

To arrive at the correct solution, we must identify the cards that have the potential to disprove the given statement. The crucial observation lies in recognising that only a card displaying 28 on one side and something other than R on the other side can invalidate the statement.

In this scenario, the cards we need to focus on are the 28 card and the B card. Let's explore the reasoning behind this.

THE CORRECT SOLUTION: MINIMUM NUMBER OF CARDS

The Card with 28 on Its Face-Up Side: This is the most direct test of the statement. If the other side is not R, the statement is false.

The Card with B on Its Face-Up Side: This card needs to be checked because if the other side is 28, it would contradict the statement. The statement only mentions what is on the other side of 28, not what is on the other side of R.

The cards with 23 and R on their face-up sides do not need to be checked. The card with 23 is irrelevant to the rule, which only concerns 28. The card with R does not need to be checked because the rule does not specify what should be on the other side of R.

So, you only need to turn over two cards: the one showing 28 and the one showing B.

CONCLUSION: BEYOND INITIAL IMPRESSIONS

The Wason selection task, or the Four Card Problem, immerses us in the depth of logical analysis and conditional reasoning. By identifying the two necessary cards to flip, the 28 and the B, we confront the task's real challenge and learn the importance of testing for falsification rather than confirmation.

The puzzle serves as a powerful reminder of the complexities that lie beneath seemingly simple tasks and the importance of careful analysis when engaging in logical problem-solving. It challenges us to think beyond initial assumptions and consider the logical implications hidden within the given information. As such, it is a clear reminder of the complexities hidden within seemingly straightforward problems and the value of meticulous analysis in navigating the world of logic.

When Should We Be Sure? A Lesson in Certainty

SURE

Life's lessons often come from the least expected places, and here's a true tale that serves as proof. If you've ever been certain of something, or think you might be in the future, this story is for you. It's titled 'A Lesson in Certainty'.

A LESSON IN CERTAINTY

It was an early spring morning that challenged my classmate Steve's philosophy for the first time. Sure, he acknowledged the certainties of mathematics—like two plus two equalling four—but real life, he believed, was a different, more unpredictable beast. Yet, on the day of the Latin mock exam, his outlook took an unexpected turn. Steve didn't see the point in these practice exams; to him, they were trivial. But not only was he proficient in Latin, he was outstanding, having even impressed the headmaster's Latin-speaking wife. And she spoke it like a centurion, or so the headmaster had once declared at morning prayers.

Confident in his rare proficiency, I was bemused when Steve wagered he'd score below 10%. I took the bet, convinced his claim was a bluff. After all, how could someone so skilled in Latin perform so poorly? When the mock began, Steve did nothing more than write his name on his paper, ensuring the outcome he predicted. Upon learning what he did, I was certain he'd secured his victory in our wager.

A week later, the scores were announced. "Steve Bryce has come bottom of the class," thundered the headmaster's wife. But now came the twist: due to the overall poor performance of the class, it was decided that the exam was overly difficult and an additional 10% was awarded to everyone, pushing Steve to exactly 10%.

I won the bet, but the lesson we both learned was invaluable. Certainty, we discovered, is a fragile concept in life's grand scheme. It was a lesson we were sure never to forget.

A Question
of Strategy

7

When Should We Expect to Beat the Market?
Exploring Some Strange Anomalies

THE HALLOWEEN EFFECT:
AN ENIGMA ROOTED IN HISTORY

The world of finance is awash with anomalies that often defy straightforward logic, but few are as intriguing and laden with history as the Halloween Effect. This peculiarity, traceable back to 1694, coincides with other pivotal events like the establishment of the Bank of England and the birth of the Enlightenment writer, Voltaire.

In their ground-breaking paper in 2002 titled 'The Halloween Indicator, Sell in May and Go Away: Another Puzzle', Sven Bouman and Ben Jacobsen tested the hypothesis that stock market returns tend to be significantly higher in the November–April period than in the May–October period. This was a claim long circulated within investment circles but never subjected to rigorous empirical analysis.

They studied 37 developed and emerging markets from 1973 to 1998 and the results were striking. The Halloween Effect was found to be valid in 36 of the 37 markets examined.

It is theorised to have begun with the landed gentry in England, who would liquidate their stock holdings in May to fund their leisurely summer sojourns to the countryside. In November, they would return to the city and reinvest their capital, thus leading to a surge in the market.

While such a practice might seem archaic in a modern context and hardly the foundation for a strategy in today's vastly more complex global markets, it's intriguing to consider that vestiges of this historical behaviour pattern might still influence market trends today.

DOI: 10.1201/9781003402862-7

More generally, the Halloween Effect is one of the more evident examples of so-called 'calendar anomalies', a term used to describe consistent and patterns of returns associated with specific calendar periods. Such anomalies pose interesting questions about market efficiency and investor behaviour.

As a trading strategy, however, it's essential to approach any such anomalies with a healthy dose of caution. If something in the world of finance looks too good to be true, it's always possible that's because it is!

THE SUPER BOWL PREDICTOR: FUSING SPORTS AND STOCKS

The Super Bowl Predictor, arguably one of the most unusual market anomalies, under-scores the idea that even the most unrelated events could potentially influence market movements. As bizarre as it sounds, this predictor connects the high-stakes world of American football to the turbulent swings of the stock market.

This peculiar theory was first propounded by sportswriter Leonard Koppett in 1978, who suggested a correlation between the outcome of the Super Bowl and the movement of the stock market for the ensuing year. The prediction hinges on the affili-ation of the winning team: in simple terms, a win by a team from the National Football Conference (NFC) signals a bull market, an up year for the stock market, whereas a win by an American Football Conference team forecasts a bear market, a down mar-ket. There are some caveats, based on the original provenance of the team, but that's the basic premise.

Robert Stovall, an influential finance expert, was an ardent supporter of the Super Bowl Predictor. He continued to popularise and champion this unconventional market indicator throughout his career, contributing to its widespread recognition.

As bizarre as this indicator might seem, it boasted a remarkable success rate of 90.3% until 1997. In other words, for nearly two decades, the outcome of a single foot-ball game purportedly offered a good indicator of the stock market's future direction. Since then, however, the Predictor has met with far less success.

Critics dismiss the Super Bowl Predictor as a product of chance. Still, its existence has spurred critical discourse about the role of randomness in markets and the search for patterns in unpredictable systems.

So, is the Super Bowl indicator a real forecasting tool? Consider the words of Fischer Black about a very different apparent anomaly: '... it sounds like people searched over thousands of rules until they found one that worked in the past. Then they reported it, as if past performance were indicative of future performance. As we might expect, in real life the rule did not work anymore'.

A SUNNY DISPOSITION: THE ATMOSPHERE OF FINANCE

An intriguing anomaly that entwines meteorology with finance is an observed correlation between weather patterns and stock market performance. This peculiarity underscores the non-financial, psychological, and even biological factors that can influence market movements, challenging the traditionally rational assumptions about market behaviour.

In 1993, Edward Saunders explored this weather-based anomaly in the American Economic Review, revealing a fascinating correlation between clear skies and elevated stock market returns. A decade later, in 2003, David Hirshleifer and Tyler Shumway bolstered this connection with their own research. They discovered that morning sunshine correlated with higher daily stock returns, a phenomenon that persisted even when accounting for other known influences on stock prices.

When the morning is sunny, it is theorised that people tend to feel more optimistic, causing a boost in their mood. This improved mood can then translate into more positive trading behaviour, potentially leading to increased stock prices. Conversely, overcast conditions could lead to more pessimistic moods, which in turn could trigger more conservative or negative trading behaviours, leading to a dip in stock prices.

While some may find this explanation of a link between weather and stock market performance somewhat outlandish, there is a range of published evidence of a statistically significant effect across an array of markets.

THE PALM PUZZLE

The 3Com-Palm puzzle is a classic example of an anomaly often discussed in behavioural finance, associated particularly with Richard Thaler.

Here's a more detailed breakdown:

1. **3Com and Palm**: 3Com was a technology company that owned a significant stake in Palm, Inc., the maker of Palm Pilot handheld devices. Palm was a highly anticipated and successful share offering at the time.
2. **Market Anomaly**: The anomaly occurred when, after the share offering, the market value of 3Com's stake in Palm (calculated by multiplying the number of shares 3Com held by the market price per share of Palm) turned out to be greater than the entire market capitalisation of 3Com itself. This was paradoxical because 3Com's valuation should logically have included the value of its Palm shares plus all other 3Com assets and operations, which suggests it should have been worth rather more than its stake in Palm alone.

3. **Behavioural Economics Significance**: This scenario contradicted traditional market valuation theories, which propose that markets are efficient and securities are priced rationally based on available information. The 3Com-Palm case highlighted market inefficiencies and became a significant example for behavioural economists, who argue that markets can be influenced by irrational and psychological factors, leading to deviations from theoretical expectations.

THE ROYAL DUTCH SHELL ANOMALY

The law of one price states in simple terms that in efficient markets, identical goods should have only one price. This principle underpins much of financial theory, which often assumes markets are efficient and participants are rational.

The case of Royal Dutch Shell is a classic example used in financial studies to illustrate market inefficiencies and the impact of investor sentiment, among other factors, on stock prices. Royal Dutch Shell was formed from the merger of Royal Dutch Petroleum of the Netherlands and Shell Transport and Trading of the UK in 1907. The companies operated under a complex structure, with shareholders receiving dividends based on a fixed ratio (60% for Royal Dutch and 40% for Shell), which theoretically should have led to a predictable and stable price ratio between the two stocks if markets were perfectly efficient and followed the law of one price.

However, in practice, the price ratio between Royal Dutch and Shell shares often deviated very significantly from the expected 1.5 (reflecting the 60/40 dividend ratio).

This case is often used to demonstrate that real-world markets are not always efficient and that prices can be influenced by a range of rational and irrational factors.

CONCLUSION: OPPORTUNITY AND CAUTION

Market anomalies such as the Halloween Effect, the Super Bowl Predictor, the Weather Effect, the Palm Puzzle, and the Royal Dutch Shell anomaly, illuminate the intricate and sometimes puzzling nature of financial markets. These anomalies challenge the Efficient Market Hypothesis (EMH), a foundational concept of finance which asserts that all relevant information is already factored into asset prices, and consistent abnormal returns are thus only achievable by chance. So, do they really serve as a key to unlocking future returns, or are they merely statistical oddities without much practical trading value?

Ultimately, prudence is paramount. There are several factors to consider before diving headlong into strategies based on these phenomena. For one thing, these anomalies are based on historical patterns, and as every seasoned investor knows, past performance is not a guaranteed indicator of future results. These patterns could change or disappear or even reverse over time due to shifts in market structure, regulation, or investor behaviour.

Broadly speaking, though, the concept of market anomalies raises a pertinent question. If we spot what seems to be a market inefficiency that we think can be leveraged for financial gain, what steps should we take to capitalise on it? A crucial consideration here is determining why other traders have not already taken advantage of this anomaly to the point where the inefficiency is neutralised. We might flatter ourselves thinking that our intelligence, diligence, or superior algorithms set us apart. While these reasons could be valid in some instances, it's more probable that we've either misunderstood the situation or that the costs of capitalising on this inefficiency—in terms of risks, potential for losses, or unexpected downsides—are too high. We should also question if the scale of the available trade justifies the time and effort involved.

Despite all this caution, if there appears to be an anomaly that's simple to exploit without apparent compensating downsides or costs, it could be a rare opportunity that is as real as it seems. Consider the analogy of the £20 note on the pavement that the financial economist dismisses as counterfeit, reasoning that if it were real, someone would have already picked it up. Occasionally, it might just be worth taking that closer look.

When Should We Believe the Crowd?
Exploring Some Lessons from a "Lost Play"

WISDOM OF THE CROWD

Are the many wiser than the few? Are the masses cleverer than the expert? This is a question which in recent years attracted an explosion of interest. The idea is often traced to a paper published in 1907 in the science journal, Nature, by Sir Francis Galton, cousin of Charles Darwin. In that paper, titled Vox Populi (Voice of the People), he demonstrated the results of a simple averaging of all the entries into a competition to guess the weight of an ox at a country fair.

He found that that the average (mean) guess was 1,197 pounds, just one pound off the actual weight of 1,198 pounds. This power of 'crowd wisdom' to outperform individual experts has been demonstrated in numerous examples since, from locating a submarine missing in the Atlantic Ocean to predicting the outcome of national elections and even papal conclaves.

Harnessing People Power

Harnessing the wisdom of the crowd, often referred to as "people power," taps into the collective intelligence of a broad spectrum of individuals to enhance decision-making, improve forecasting accuracy, and solve complex problems more efficiently than could be achieved by individual experts or smaller groups. This approach is founded on the principle that a diverse array of people, each bringing their own unique information, perspectives, and insights, can contribute to a more accurate and insightful outcome when their collective knowledge is aggregated.

A quintessential manifestation of this concept is found in prediction markets. These platforms operate similarly to financial markets, allowing participants to buy and sell contracts based on their predictions about future events. The pricing of these contracts reflects the crowd's collective assessment of the likelihood of various outcomes, leveraging the same underlying principle that the value of financial assets incorporates all known information about future earnings and associated risks.

Prediction markets have found application across a broad range of fields, demonstrating their versatility and utility:

- **Public Health**: Used to forecast the trajectory of infectious diseases, these markets can inform public health strategies and resource allocation, aiming to mitigate the effects of potential outbreaks.
- **Healthcare Demand**: By predicting the demand for hospital services, prediction markets can enable more effective resource management, staff allocation, and operational efficiency, ultimately enhancing patient care.
- **Entertainment Industry**: In the realm of cinema, these markets can provide early insights into potential box office performance, guiding studios and distributors in their marketing and strategic decisions.
- **Project Management**: Organisations employ prediction markets to assess the probability of meeting project deadlines, facilitating early identification of possible delays or problems for timely corrective actions.
- **Public Policy**: Insights derived from prediction markets extend to public policy, aiding in the anticipation of election results, economic trends, and the effects of regulatory changes, thereby supporting more informed policymaking.

The power of prediction markets lies in their capacity to synthesise diverse opinions and information, mitigating individual biases and errors, often resulting in forecasts that surpass the accuracy of expert predictions or traditional forecasting methods. However, the effectiveness of these markets hinges on several critical factors, including the diversity and independence of participants, market liquidity, and the design of market mechanisms that accurately reflect collective beliefs.

Thus, while prediction markets present significant opportunities for harnessing "people power," their deployment must be judiciously managed to unlock their full potential. They represent a shift towards more democratic, collective approaches in forecasting and decision-making, where every individual's perspective is valued, leading to more precise and actionable collective insights across various domains.

Vortigern and Rowena

It's hard to beat the fascination of watching the power of crowd wisdom when it's demonstrated in the raw, when an actual crowd demonstrates its dominance over the experts.

We can go back a lot further than the days of Galton's ox for a powerful example, and it involves a late 18th-century tale of William Shakespeare and a 'lost play'. The play was called Vortigern and Rowena and was widely proclaimed as a lost work of the Bard.

It was championed by James Boswell, the acclaimed diarist and biographer of Samuel Johnson, by Henry James Pye, the poet laureate, and by the playwright Richard Brinsley Sheridan. To widespread delight, Vortigern and Rowena opened to a packed, enthusiastic audience on the evening of April 2, 1796. The part of Vortigern himself was played by no less a light than the acclaimed Shakespearean actor, John Philip Kemble.

The widespread excitement and anticipation among the audience soon turned to bemusement, however, and then literal disbelief, so that by the time Kemble was drawn to hint at his own opinion, repeating with emphasis Vortigern's line "and when this solemn mockery is o'er", the catcalls of the audience told their own story. One performance before a crowd of ordinary theatregoers was enough to kill off any notion that this was a genuine work of the Bard of Avon. The real author, William Henry Ireland, soon admitted to the hoax and promptly left for France.

A Lesson Learned

So, what can we learn here about the wisdom of crowds? Is it perhaps the case that Shakespeare is to be played, not read, and the 18th-century experts who examined it simply took it on trust that it would appear better when played than read? Could it be that the real experts were the performers who had played much of the canon of the authentic William Shakespeare – and that their sceptical performances tipped the wink to the theatregoing crowd? Or could it be that the crowd simply was as wise as its reaction suggests? Whatever is the case, of something we can be sure. One crowd of paying spectators was enough. Vortigern and Rowena didn't open for a second day.

When Should We Cooperate? Exploring the Elements of Game Theory

INTRODUCING GAME THEORY

The formal foundation for modern game theory was laid by John von Neumann and Oskar Morgenstern in the 1940s, with their ground-breaking book, *Theory of Games and Economic Behavior*. This work sets the stage for game theory as a crucial analytical tool in economics. In the 1950s, John Nash extended its scope, introducing concepts like the Nash Equilibrium. These milestones marked game theory's evolution from a mathematical curiosity to a pivotal tool in various disciplines.

Essentially, game theory focuses on examining how rational individuals (or entities, which could for example be groups, organisations, or countries) make decisions to maximise their own outcomes in situations where the decisions of one so-called 'player' affect the outcomes of other 'players', and vice versa. It provides a framework to analyse and predict the choices of rational actors in these strategic situations.

APPLICATIONS IN VARIOUS FIELDS

The versatility of game theory is evident in its wide-ranging applications across multiple fields. It can be applied to a wide range of scenarios, from everyday decision-making and business strategy to complex negotiations and interactions in international relations. In economics, it models market dynamics, helping understand competitive strategies and the dynamics of auctions. Biologists use game theory to analyse animal behaviour, such as mating rituals and foraging strategies, viewing them as strategic games for survival. In political science, game theory is applied to electoral strategies, voting systems, and international diplomacy, providing insights into the strategic behaviour of voters, politicians, and nations.

UNDERSTANDING THE NASH EQUILIBRIUM

An essential concept in game theory is the Nash Equilibrium. This represents a state in which, given the strategies of the other players, each player's chosen strategy maximises their payoff and they have no incentive to deviate from it. In other words, a Nash Equilibrium occurs when all players choose the best response to the other players' strategies.

In a game reaching a Nash Equilibrium, no player can unilaterally improve their situation by changing their strategy, assuming other players' strategies remain fixed.

Take the case of two spies, Anna and Barbara: if they both use the same code, they successfully communicate and gain a reward; if they use different codes, the communication fails, resulting in no payoff. Here, the Nash Equilibria are the scenarios where both spies choose the same code—either code 1 or code 2. In these situations, no player can improve her payoff by unilaterally changing her strategy given the strategy of the other.

Similarly, two drivers must decide which side of the road to drive on. Assuming that their concern is to avoid a collision, the Nash Equilibria occur when both drivers choose to drive on the same side, either left or right. It is beneficial for each driver to imitate the choice of the other; if one driver deviates, it would result in a collision. In this case, therefore, there are two Nash Equilibria—both drive on the left or both drive on the right.

In other situations, no stable state or Nash Equilibrium can be achieved. Take as an example two companies choosing a company logo. If Company A chooses a bear logo,

Company B may have an incentive to switch its own logo to a bull. However, if it does so, Company A may now have an incentive to change its logo again, and so on. This situation has no Nash Equilibrium, where a Nash Equilibrium means no company would benefit by changing its strategy given the strategy of the other.

These examples show how the concept of the Nash Equilibrium is applied in different contexts to predict and analyse strategic choices. However, they also underscore the complexities of game theory and strategic interactions. Some games might not have a Nash Equilibrium, and in others, there may be multiple equilibria.

The concept of the Nash Equilibrium becomes more interesting and complex in larger games with more players and more possible strategies. For instance, in the marketplace, firms' pricing strategies could reach a Nash Equilibrium where no single firm could increase its profit by unilaterally changing its price, given the prices of its competitors. An example is a simple two-firm scenario where both firms know that if either raised their price, they would lose most or all of their customers to the other firm.

However, it's important to note that Nash Equilibria do not always lead to the best collective outcome. The classic example is the 'Prisoner's Dilemma', where the Nash Equilibrium strategy for both players leads to a worse collective outcome than if they could collaborate. This highlights that while the Nash Equilibrium is a powerful concept in understanding strategic behaviour, it is not necessarily synonymous with individual or group optimality.

UNDERSTANDING THE PRISONER'S DILEMMA

The Prisoner's Dilemma is a classic problem in game theory, illustrating why it can be hard for rational individuals to cooperate, even when it is in their best interest. In this scenario, two prisoners are individually given the option to confess or deny a crime they have committed together. Depending on the combination of their decisions, they can either reduce their sentences, remain the same, or one can go free while the other gets a heavier sentence. They cannot communicate or collude.

If one confesses and the other doesn't, the prisoner who confesses is released. If both confess, each is better off than denying while the other confesses. The Nash Equilibrium in this game is for both prisoners to confess, which is not the optimal outcome for either. This problem illustrates a situation where individuals' rational decisions can lead to a collectively undesirable outcome.

Let's demonstrate this with an example:

Imagine two prisoners who are part of the same crime. They can't talk to each other. Here is their dilemma:

- If both confess, they each get two years in jail.
- If one confesses and the other denies, the one who confesses goes free and the other gets eight years.
- If both deny, they each get only one year in jail.

The smartest move for two self-interested prisoners is in each case to confess, because they can't be sure what the other will do. This is the Nash Equilibrium. But if they could make a deal, they would both deny the crime and get just one year each.

This situation also shows a 'dominant strategy', where the best choice (confessing) doesn't depend on what the other person does. It's the best move no matter what.

But not all situations have a dominant strategy. Take driving on the right or left side of the road. In the US, driving on the right is the norm, so it's best for everyone to do that. In the UK, it's driving on the left. These are examples of Nash Equilibria, where everyone's doing what is best considering what others are doing.

So, a Nash Equilibrium is a stable situation where nobody gains by changing their strategy if others don't change theirs. It's not always the best for everyone involved, but it's often what happens, especially among rational, self-interested people. Sometimes the best strategy in theory is not the best in practice.

GOLDEN BALLS DILEMMA

An example of the Prisoner's Dilemma in action is a one-time TV show called 'Golden Balls' where two players each choose a ball—either 'Split' or 'Steal', without knowing what the other chooses. They can talk before choosing. Here's what happens next:

- If both choose 'Split', they share the prize money equally.
- If both choose 'Steal', neither gets any money.
- If one chooses 'Steal' and the other 'Split', the 'Steal' player gets all the money, and the 'Split' player gets nothing.

In this game, the best bet for self-interested players (as in the Prisoner's Dilemma) is to both choose 'Steal', because choosing 'Steal' is always better or no worse than choosing 'Split'. 'Steal' in this game is like 'Confess' in the Prisoner's Dilemma.

The difference is that in the Prisoner's Dilemma, the players can't talk to each other. In Golden Balls, they can. They could both win half the prize if they agree to 'Split', but they risk losing everything if they both choose 'Steal'. The show often has players agreeing to 'Split', but then one or both betray the agreement and pick 'Steal'.

This demonstrates that even when players can talk and agree in games like these, they can still end up not cooperating if there's no way to enforce their agreement. Indeed, the more credible is the promise to split, the more tempting it may be for the opponent to steal. This tells us that not even communication and agreement can resolve the Prisoner's Dilemma when there is no way to enforce the agreement. This 'problem of credible commitment' is a common feature of many strategic interactions in real life.

TRAITOR'S DILEMMA

In the reality TV show "Traitors" there is a variant of the Golden Balls Dilemma but with more players, called the Traitor's Dilemma.

To analyse this, we can consider the different scenarios and the rewards associated with each. The outcomes depend in this example on the decisions made by three players:

1. If all decide to share, they each get one-third of the pot.
2. If all decide to steal, they all get nothing.
3. If two decide to steal and one decides to share, the two who chose to steal split the pot (each getting half).
4. If one decides to steal and the others decide to share, the one who chose to steal takes the whole pot.

First, notice that if a player expects the other two to Share, their best response is to Steal, since this would give them the entire pot instead of just one third. If a player expects one to Steal and one to Share, their best response is also to Steal, ensuring they get at least half the pot (if two steal) instead of nothing. If a player expects both others to Steal, their best strategy is indifferent between Stealing and Sharing since both result in no prize; however, in typical game-theoretical analysis, such players might lean towards Stealing out of self-interest, as it does not worsen their situation but has a potential benefit.

Thus, we can infer that in each situation, choosing to Steal can never result in a worse outcome for a player than choosing to Share, assuming the other players' actions are fixed.

However, this equilibrium is precarious in real-world contexts, especially if the players can communicate or have formed trust throughout the game, as mutual cooperation (all choosing to Share) leads to a better outcome for the group compared to the individual rationality of Stealing.

So, in the context of game theory, the optimal strategy in a single shot of this game, without considering trust or external factors, would be to Steal, as it maximises the player's minimum possible gain, given the assumptions typical in game theory of rationality and self-interest. However, "optimal" can vary based on the context of previous rounds, relationships, or possible future repercussions outside the standard game-theoretical framework.

ITERATED GAMES AND REPUTATION

One solution to the problem of credible commitment is through the concept of iterated games and the development of a reputation. An iterated game is a repeated version of a basic game. In these games, players can observe the actions of their opponents over multiple rounds, allowing them to adjust their strategies based on their opponents' past behaviour.

In the context of our 'Golden Balls' or 'Traitors' examples, if the game were to be played repeatedly a player who breaks their promises would quickly gain a reputation for being untrustworthy. Knowing this, they would be more likely to keep their promises to maintain their reputation and the trust of their opponents.

THE DOLLAR AUCTION

The 'Dollar Auction' paradox relates to a scenario where Mr. Moneymaker auctions off dollar bills. The rule is that the highest bidder wins the dollar, but the second-highest bidder also must pay their bid and gets nothing. Here's an example:

- Someone bids 1 cent, hoping to make 99 cents profit.
- Then another bids 2 cents, and so on, up to 99 cents.
- At 99 cents, the person who bid 98 cents doesn't want to lose that money, so they bid $1.
- This keeps going, with each bidder trying to avoid losing their bid amount. It becomes a cycle where the only winner is Mr. Moneymaker, the auctioneer.

This example, along with the previous ones, highlights the successes and failures of communication and coordination. Finally, let's touch on 'focal points' or 'Schelling points'. These are strategies people naturally pick to coordinate without communication. An example is when people were asked to meet a stranger in New York City without any specific instructions or prior communication. Many chose 12 noon at Grand Central Station as the meeting point. This is a 'Schelling point' because it's a natural and obvious choice for coordination in the absence of communication.

CONCLUSION: GAME THEORY— A POWERFUL TOOL

The Prisoner's Dilemma, Nash Equilibrium, and the broader field of game theory provide powerful tools for analysing situations of strategic interaction. While these tools can highlight potential outcomes and strategies, they also expose the inherent challenges involved in these situations, such as the problem of credible commitment and the importance of reputation in iterated games. By understanding these concepts, we can better understand the complex dynamics of many real-world scenarios, from business negotiations to international politics.

When Should We Forgive and Forget?
Exploring Repeated Game Strategies

REPEATED GAMES

The Prisoner's Dilemma exemplifies a one-stage game, where there is no repercussion or continuation after a player chooses to confess or deny and the interplay ends. This is obviously not representative of most real-world scenarios, which often involve multi-stage interactions and decisions that are influenced by previous outcomes. This leads us to the realm of repeated games.

UNCERTAINTY AND REPEATED GAMES

In many real-world situations, it's often unclear when the game will end, a scenario that can be modelled by rolling two dice after each round in a game. If a double-six is rolled, the game ends. For any other combination, you play another round, with the game continuing until a double-six is rolled. Your total score for the game is the sum of your payoffs.

THE SEVEN PROPOSED STRATEGIES
IN REPEATED GAMES

In such games of repeated rounds with no defined end-point, several strategies emerge, which we can model for simplicity by assuming that there are two possible decisions for each player at every stage: to cooperate and 'split' in a Golden Balls game-style scenario (be friendly), or to be selfish ('steal' in a Golden Balls game-style scenario). In a repeated game, we can model this friendly/hostile choice into seven scenarios:

1. **Always Friendly:** This strategy involves being friendly every time.
2. **Always Hostile:** As the name suggests, this strategy involves being hostile every time.

3. **Retaliation:** This strategy requires you to be friendly as long as your opponent is friendly, but if your opponent is ever hostile, you should turn hostile.

4. **Tit for Tat:** This strategy starts with being friendly and then replicating your opponent's previous move in subsequent rounds.

5. **Random Approach:** This strategy suggests tossing a coin for each move and deciding based on the outcome.

6. **Alternate:** This strategy alternates between being friendly and hostile.

7. **Fraction:** This strategy involves starting friendly and remaining so if the fraction of times your opponent has been friendly is more than half.

DETAILED ANALYSIS OF THE SEVEN PROPOSED STRATEGIES IN REPEATED GAMES

Understanding the dynamics of indefinite repeated games often involves exploring various strategies that players can adopt. Let's go deeper into the seven strategies outlined:

1. **Always Friendly:** Here, the player adopts a cooperative approach, choosing to be friendly in every round. This strategy could lead to high payoffs when interacting with other friendly players but leaves the player vulnerable to exploitation by hostile players.

2. **Always Hostile:** This strategy is the opposite of the 'Always Friendly' approach. The player chooses to be hostile in every round, aiming to exploit friendly opponents. However, when encountering another hostile player or retaliatory strategies, the outcome can be less favourable.

3. **Retaliation:** The player starts friendly and remains so if the opponent does the same. However, if the opponent ever chooses to be hostile, the player shifts to a permanently hostile stance. This strategy can deter hostile behaviour from the opponent but might lead to an endless cycle of hostility if triggered.

4. **Tit for Tat:** This strategy is famous for its effectiveness and simplicity. The player starts friendly and then mimics the opponent's behaviour from the previous round. It rewards cooperation and retaliates against hostility, but it also forgives after retaliation since it reverts to cooperation if the opponent does so.

5. **Random Approach:** The player's choice of action is determined by a coin toss, making this strategy completely unpredictable. While this randomness might confuse the opponent, it also disconnects the player's actions from the opponent's behaviour, making it less effective in promoting cooperation.

6. **Alternate:** The player alternates between friendly and hostile actions. Again, this does not adapt to the opponent's behaviour and thus may not be an optimal strategy.

7. **Fraction:** This strategy starts friendly and then assesses the overall behaviour of the opponent. If the opponent has been friendly more than half of the

time, the player continues to be friendly; otherwise, they turn hostile. This strategy attempts to mirror the opponent's overall conduct but might be less responsive to recent changes in behaviour.

DOMINANT STRATEGY IN REPEATED GAMES

Although there's no dominant strategy in such games, simulations of tournaments where each strategy plays against every other have often shown Tit for Tat to emerge victorious. This strategy tends to win overall because it performs well against friendly strategies without being exploitable by hostile ones. The key attributes contributing to the success of Tit for Tat are its niceness, retaliation, forgiveness, and clarity.

A DEEPER DIVE INTO THE SUCCESS OF 'TIT FOR TAT' IN REPEATED GAMES

Repeated games offer an intricate canvas for strategic interactions. Although no strategy dominates all others universally in such scenarios, Tit for Tat often proves to be the most successful one overall in a variety of conditions. This effectiveness results from several of its unique characteristics:

1. **Niceness:** A Tit for Tat player starts off by being friendly or cooperative. By not being the first to defect, it encourages cooperative behaviour right from the start. It shows goodwill to its opponents, promoting an atmosphere of trust and mutual benefit.
2. **Retaliation:** Tit for Tat is not naive; it does not allow exploitation. If an opponent chooses to defect or act hostile, Tit for Tat will retaliate in kind in the next round. This principle of immediate retaliation provides a clear disincentive for opponents to defect, knowing that such behaviour will not go unpunished.
3. **Forgiveness:** Despite its willingness to retaliate, Tit for Tat is also forgiving. If an opponent returns to cooperative behaviour after a round of defection, Tit for Tat will reciprocate with cooperation in the next round. This characteristic allows for the possibility of restoring cooperation, even after bouts of hostility.
4. **Clarity:** Tit for Tat is an easy strategy to understand and predict. It does not use complicated rules or random behaviour. This clarity makes it easier for opponents to comprehend and adapt to Tit for Tat, encouraging long-term cooperation.

Tit for Tat also provides valuable insights beyond game theory. Its fundamental principles—niceness, retaliation, forgiveness, and clarity—are effective guidelines for a wide range of social, economic, and political interactions. They capture the essence of how to balance cooperation and self-defence, trust and scepticism, and how to promote stable and beneficial relationships even in a world of self-interested individuals.

REAL-WORLD APPLICATIONS AND EXAMPLES

In international relations, strategies like Tit for Tat are evident in diplomatic negotiations and trade agreements, where countries often reciprocate actions (both positive and negative). In the corporate world, companies frequently use a mix of cooperative and competitive strategies based on their competitors' actions. Environmental agreements often see a blend of these strategies, where nations commit to certain standards and adjust their policies in response to the actions of others.

PSYCHOLOGICAL AND SOCIOLOGICAL ASPECTS

The success of strategies like Tit for Tat in repeated games reflects certain psychological and sociological truths. For instance, the effectiveness of Tit for Tat aligns with psychological principles of reciprocity and fairness, suggesting an innate human tendency to respond to cooperation with cooperation. Sociologically, these strategies highlight the importance of norms and trust in maintaining cooperative relationships within societies.

LIMITATIONS AND CRITICISMS

While strategies like Tit for Tat have been celebrated for their simplicity and effectiveness, they are not without limitations. In complex real-world situations, including those involving multiple players with varying objectives, these strategies can sometimes lead to suboptimal outcomes. For example, Tit for Tat can lead to endless cycles of retaliation in certain situations.

EMERGING TRENDS AND FUTURE RESEARCH

With the advent of artificial intelligence and machine learning, new strategies in repeated games are emerging. These technologies allow for the analysis of vast amounts of data to identify patterns and develop strategies that were previously too complex to model. Research is also focusing on how these strategies can be adapted in the digital world, particularly in areas like cybersecurity, where strategic interactions are frequent and complex.

CONCLUSION: THE EVOLUTION OF COOPERATION

The success of the Tit for Tat strategy in repeated games isn't an isolated phenomenon. Whether in interpersonal relationships, business negotiations, or global diplomacy, the lessons from Tit for Tat are valuable, timeless, and universal. Robert Axelrod, in his book *The Evolution of Cooperation*, commends it for its willingness to retaliate tempered by forgiveness. He also commends its clarity and essential niceness. These principles, when applied in real life, can lead to more harmonious and successful interactions, paving the way for positive outcomes in various situations. Even so, the strategy, especially in a complex and interconnected world, is not without its limitations.

When Should We Mix It Up? Exploring Mixed Strategy Methods in Game Theory

INTRODUCTION

In game theory, the concept of mixed strategy arises when players face a decision-making situation where they do not have a dominant strategy. A dominant strategy is a strategy that is always better than any other strategy, regardless of the opponent's choice. However, in some cases, players employ a mixed strategy, which involves randomising their choices to maximise their expected payoffs.

CREATING UNCERTAINTY

The purpose of employing a mixed strategy is to create a balanced approach that maximises expected payoffs. By introducing randomness into their decision-making, players can create uncertainty for their opponents and avoid predictability. This uncertainty makes it difficult for opponents to exploit any patterns or weaknesses in the player's choices.

In situations where no dominant strategy exists, players use mixed strategies to find a Nash Equilibrium, which is a state where no player can improve their payoff by unilaterally deviating from their current strategy. Nash Equilibria often involve players randomising their choices, as this creates a balance where no player can gain an advantage by deviating from their strategy.

Mixed strategies offer a powerful tool for players to navigate complex strategic interactions. By incorporating randomness, players can mitigate the risk of being exploited by opponents who attempt to exploit predictable behaviour. Instead, mixed strategies introduce a level of unpredictability, making it challenging for opponents to determine the player's intentions and respond optimally.

ROCK-PAPER-SCISSORS

To illustrate this, consider a simple example of a two-player game, such as Rock-Paper-Scissors. In this game, each player has three pure strategies: Rock, Paper, and Scissors. If one player randomises their choices by assigning equal probabilities to each strategy, they introduce uncertainty into the game.

For instance, Player A might choose to play Rock, Paper, or Scissors with equal probabilities of 1/3 each. In response, Player B might also choose to play Rock, Paper, or Scissors with equal probabilities of 1/3 each.

In a real-life high stakes environment, the strategy has in fact been different, at least in one high profile case. I refer to the year 2005, when the president of Japanese electronics giant Maspro Denkoh Corporation faced a significant dilemma regarding the auction of the company's prestigious art collection. Valued at around $20 million, the decision of whether Christie's or Sotheby's, both historic auction houses, should handle the auction was a challenging one. Unable to decide, he resorted to the game of rock, paper, scissors. This choice was seen as a fair way to resolve the impasse between the two firms.

Christie's approach to the challenge was meticulous; the president of Christie's in Japan researched the psychology behind the game and even consulted children who suggested avoiding 'rock' as the initial throw due to its predictability. Their strategy was to start with 'scissors'. This move relied on the idea that Sotheby's would anticipate a 'rock' throw from Christie's and thus choose 'paper' to counteract it.

If the game had been structured as a best-of-three, Christie's could have adapted their strategy based on findings from the State Key Laboratory of Theoretical Physics in China, which suggest that winners do not randomise but in practice tend to stick with their winning choice in the subsequent round.

However, this single-round match left no room for redemption or strategic evolution. Representatives from both auction houses met at Maspro's offices, where they wrote down their selections. Christie's emerged victorious with 'scissors', defeating Sotheby's 'paper'. This led to Christie's winning the right to auction off the Maspro collection.

The auction, aptly nicknamed "Scissors", culminated in the sale of several important works, including one of Cézanne's paintings, which alone sold for $11.7 million at Christie's New York salesroom!

Back now to the world of imagination and consider a penalty awarded during a championship final.

THE PENALTY

In the 88th minute of the match, a penalty is awarded against the defending champions. The penalty taker in our simplified scenario has two options: aim straight or aim at a corner. Similarly, the goalkeeper has two choices: stand still or dive to a corner. The probabilities of scoring or saving the penalty are as follows.

PENALTY TAKER'S PROBABILITY OF SCORING

Aims Straight/Goalkeeper Stands Still: 30% chance of scoring.
Aims Straight/Goalkeeper Dives: 90% chance of scoring.
Aims at Corner/Goalkeeper Stands Still: 80% chance of scoring.
Aims at Corner/Goalkeeper Dives: 50% chance of scoring.

GOALKEEPER'S PROBABILITY OF SAVING

Stands still/Penalty taker aims straight: 70% chance of saving.
Stands still/Penalty taker aims at corner: 20% chance of saving.
Dives/Penalty taker aims straight: 10% chance of saving.
Dives/Penalty taker aims at corner: 50% chance of saving.

ABSENCE OF DOMINANT STRATEGIES

Neither the penalty taker nor the goalkeeper has a dominant strategy in this game. A dominant strategy would be a choice that is superior to any other strategy, regardless of the opponent's choice. Since this is not the case, both players must consider the opponent's strategy when deciding their own.

MIXED STRATEGY EQUILIBRIUM

Game theory suggests that in the absence of dominant strategies, players should adopt mixed strategies to maximise their expected payoffs. A mixed strategy involves randomising the choices according to specific probabilities.

For the penalty taker, the optimal mixed strategy in this scenario involves aiming for the corner with a two-thirds (2/3) probability and shooting straight with a one-third (1/3) probability. This ratio can be derived with a bit of algebra by finding the ratio where the chances of scoring are the same, regardless of the goalkeeper's strategy.

Likewise, the goalkeeper's optimal mixed strategy involves diving for the corner with a five-ninths (5/9) probability and standing still with a four-ninths (4/9) probability. This ratio ensures that the chance of saving the penalty is equal, regardless of the penalty taker's choice.

IMPLEMENTATION CHALLENGES

To effectively employ a mixed strategy, it is essential to introduce an element of randomness into the decision-making process. In the context of a penalty shootout, this requires the ability to randomise choices effectively.

For example, the penalty taker could use a method such as noting the time on the match clock, having divided these up mentally into six sections. If it shows section 1, 2, 3, or 4, the penalty taker aims for the corner; if it's section 5 or 6, they shoot straight. Anyway, you get the general idea. This approach ensures that the penalty taker maintains the desired probability (2/3 in this case) of aiming for the corner.

SUPPORTING STUDIES

Game theory suggests that goalkeepers should randomise in some way their dive direction to optimally counteract any choice by the penalty taker. This concept, a fundamental part of game theory, aligns with real-world findings published in the American Economic Review. Additionally, scholarly debates have explored the pros and cons of various shooting and diving strategies in soccer. For instance, a paper published in Psychological Science indicated a tendency for goalkeepers to dive more frequently to the right when their team was trailing. However, this pattern wasn't observed when their team was leading or the score was tied. Non-random patterns have also been identified in tennis, including published evidence that even professional players tend to alternate serves too regularly, while the stage of the game was again something of a predictor.

These studies provide insights into how mixed strategies and deviations from randomness can impact outcomes and shed light on the behaviour of players in high-stakes decision-making situations.

COMPLEX SCENARIOS INVOLVING MULTIPLE PLAYERS

In complex scenarios involving multiple players, such as in corporate marketing, mixed strategies become even more crucial. Companies often employ mixed strategies in competitive pricing, product launches, or market entries to avoid predictability that competitors could exploit. For instance, a company might randomise the timing of its product launches or sales promotions to keep competitors off-balance. In a multi-player market, this unpredictability can be a significant advantage, as it complicates the decision-making process for competitors.

PRACTICAL APPLICATION OF MIXED STRATEGIES

- **Military Tactics:** In military operations, mixed strategies can be employed to make it difficult for the enemy to anticipate and prepare.
- **Political Campaigns:** Political strategists often use mixed strategies in campaign messaging and policy announcements to keep opponents and voters engaged and guessing.
- **Corporate Negotiations:** Companies may use mixed strategies in negotiation tactics, alternating between hard-line and conciliatory approaches.

CONCLUSION: BACK TO THE GAME

In conclusion, game theory and the concept of mixed strategies offer valuable insights into decision-making scenarios where dominant strategies are absent. By employing randomised choices, players can maximise their expected probabilities of success. While randomising may seem counterintuitive, the application of game theory and empirical evidence from the literature demonstrate its effectiveness in real-world scenarios.

Back to the game. The shot was aimed at the left corner; the goalkeeper guessed correctly and got an outstretched hand to it, pushing it back into play, only to concede a goal on the rebound. Real Madrid got a chance to equalise from the spot eight minutes later and took it. And that's how it ended at the Bernabeu. Real Madrid 1 Barcelona 1. Honours even!

When Should We Want to Be Last?
Exploring Sequence Biases

THE CELEBRITY TALENT CONTEST

An actor, a singer, a presenter, a reality star, a comedian, a tennis player, and an assortment of other vaguely familiar faces, line up to compete for the title of best celebrity dancer. This is the well-established format of what is called 'Strictly Come Dancing' in the UK or 'Dancing with the Stars' in the US. The prize is the coveted glitterball trophy.

But how much of their success in the competition is to do with their Waltz, Foxtrot, and Charleston, and how much is it literally down to the luck of the draw?

A study published in 2010 by Lionel and Katie Page looked at public voting at the end of episodes of a singing talent contest and found that singers who appeared later in the running order received a significantly higher share of the public vote than those who had preceded them.

This was explained as a 'recency effect' meaning that those performing later are more recent in the memory of people who were voting. Interestingly, a different study, of wine tasting, suggested that there is in that arena a significant 'primacy effect' which favours the wines that people taste first (as well, to some extent, as last).

Testing for Bias

What would happen if the evaluation of each performance was carried out immediately after each performance instead of at the end? Surely this would eliminate the benefit of going last as there would be equal recency in each case? The problem in implementing this is that the public need to see all the performers before they can choose which of them deserves their vote.

In addition to the public vote, however, Strictly Come Dancing (or Dancing with the Stars in the US) includes a score awarded by a panel of expert judges immediately after each performance. There should in theory be no recency effect in this expert evaluation – because the next performer does not take to the stage until the previous performer has been scored, and so there is no 'last dance' advantage in the expert scores.

I decided to look at this using a large data set of every performance ever danced on the UK and US versions of the show – going right back to the debut show in 2004. The findings, published with two co-authors in the journal, Economics Letters, proved very surprising and counter-intuitive.

Last Shall be First

Contrary to expectations, we found the same sequence order bias by the expert panel judges – who voted after each act - as by the general public, who voted after all performances had concluded.

We applied a range of statistical tests to allow for the difference in quality of the various performers and as a result we were able to exclude quality as a reason for the observed effect. This worked for all but the opening spot of the night, which we found was generally filled by one of the better performers.

So the findings matched the 2010 study in demonstrating that the last performance slot should be most prized, but we also found that the first to perform also scored better than expected. This resembles a J-curve where the first and later performing contestants disproportionately gained higher expert panel scores. You certainly don't want to go second!

Although we believe the production team's choice of opening performance may play a role in the first performer effect, our best explanation of the key sequence biases is as a type of 'grade inflation' in the expert panel's scoring. In particular, we interpret the 'order' effect as deriving from studio audience pressure – a little like the published evidence of unconscious bias exhibited by referees in response to spectator pressure. The influence on the judges of increasing studio acclaim and euphoria as the contest progresses to a conclusion is likely to be further exacerbated by the proximity of the judges to the audience.

When the votes from the general public are used to augment the expert panel scores, the biases observed in the expert panel scores are amplified.

In summary, the best place to perform is last and second is the least successful place to perform.

The implications of this are worrying if they spill over into the real world. Is there an advantage in going last (or first) into the interview room for a job – even if the applicants are evaluated between interviews? What about the order in which your examination script appears in the pile that is being marked?

Hungry Judge Effect

A related study, published in the Proceedings of the National Academy of Science, found that experienced parole judges granted freedom about 65% of the time to the first prisoner to appear before them on a given day, and the first after lunch – but to almost nobody towards the end of a morning session. The paper speculates that breaks may serve to replenish mental resources by providing "rest, improving mood or by increasing glucose levels in the body". It's also been termed the 'hungry judge effect'. Linked to this is the concept of decision fatigue, the idea that decision-making and good judgment declines in the wake of making too many decisions without a break.

So the research confirms what has long been suspected – that the order in which things happen can make a big difference. Combined with decision fatigue there are clear implications for everyday strategy, whenever you have a choice in the matter - such as when to make that appointment with the dentist or doctor, or when to ask for a pay rise or even a date!

CONCLUSION: LEARNING SOME LESSONS

If you learn just one thing from this book, it's that life is not always about what you do, or even how you do it, but when you do it. Now think about that appointment with the dentist. Do you really want to be last in before lunch? Consider the 'hungry judge effect' and apply it to the dentist and add a touch of decision fatigue into the equation. What's your answer?

Hopefully you will have learned a lot more than this one tip from the book, but as a tip it is probably up there with the big ones! The bigger story is that there really is a lot we can learn from published research that can improve our health, happiness, and everyday lives. It's just a matter of knowing where to look and applying the lessons. Besides, it can be a whole lot of fun!

Reading and References

Aczel, A.D. (2016). *Chance: A Guide to Gambling, Love, the Stock Market and Just About Everything Else*. New York: Thunder's Mouth Press.

Adler, D. (2020). The Great Filter: A Possible Solution to the Fermi Paradox. *Astronomy*. November 20.

Albert, D. (2012). On the Origin of Everything. *Sunday Book Review, The New York Times*, March 23.

Ananthaswamy, A. (2020). Do We Live in a Simulation? Chances Are About 50-50. New York: Scientific American. October 13.

Axelrod, R. (1984). The Evolution of Cooperation. New York: Basic Books.

Axelrod, R. (2006). The Evolution of Cooperation, Revised Edn. New York: Perseus Books Group.

Axelrod, R. and Hamilton, W.D. (1981). Evolution of Cooperation, *Science*, 211, 1390–1396.

Baker, R.D. (2002). Probability Paradoxes: An Improbable Journey to the End of the World. *Mathematics Today*, December 185–189.

Barberis, N.C. (2013). Thirty Years of Prospect Theory in Economics: A Review and Assessment, *Journal of Economic Perspectives*, 27, 1, 173–196.

Barrow, J.D. (2008). *100 Essential Things You Didn't Know You Didn't Know*. London: The Bodley Head.

Barrow, J.D. (2012). *100 Essential Things You Didn't Know about Sport*. London: The Bodley Head.

Bayes, T. and Price, R. (1763). An Essay towards Solving a Problem in the Doctrine of Chances. By the Late Rev. Mr. Bayes, Communicated by Mr. Price, in a Letter to John Canton, M.A. and F.R.S. *Philosophical Transactions of the Royal Society of London*, 53, 370–418.

Bedwell, M. (2015). Slow Thinking and Deep Learning: Tversky and Kahneman's Cabs, *Global Journal of Human-Social Science*, 15, 12.

Benford, F. (1938). The Law of Anomalous Numbers, *Proceedings of the American Philosophical Society*, 78, 4, 551–572.

Bertrand, J. (1889). *Calcul des Probabilités*. Paris: Gauthier-Villars et Fils.

Bi, Z. and Zhou, H-J. (2014). Optimal Cooperation Trap Strategies for the Iterated Rock-Paper Scissors Game. PLOS ONE. October 29.

Black, F. (1992). Beating the Market: Yes, It Can Be Done. *Economist*, December 5, 23–26.

Blackburn, S. (ed.). (2016). *Ship of Theseus. The Oxford Dictionary of Philosophy*, 3rd. Edn. Oxford: Oxford University Press.

Bosch-Domenech, A., Montalvo, J.G., Nagel, R. and Satorra, A. (2002). One, Two, (Three), Infinity, … : Newspaper and Lab Beauty-Contest Experiments, *American Economic Review*, 92, 5, 1687–1701.

Bostrom, N. (2000). Cars in the Next Lane Really Do Go Faster. *+plus*, December 1.

Bostrom, N. (2003). Are You Living in a Computer Simulation? *Philosophical Quarterly*, 53, 211, 243–255.

Bostrom, N. (2006). Do We Live in a Computer Simulation? Nick Bostrom. *New Scientist*. November, 8–9.

Boumen, S. and Jacobsen, B. (2002). The Halloween Indicator: "Sell in May and Go Away": Another Puzzle, *American Economic Review*, 92, 5, 1618–1635.

Brown, C. (2005). *Aquinas and the Ship of Theseus: Solving Puzzles about Material Objects*. London: A&C Black.

Brown, A., Reade, J. and Vaughan Williams, L. (2019). When Are Prediction Market Prices Most Informative? *International Journal of Forecasting*, 35, 1, 420–428.

Brumfiel, G. (2008). The Testosterone of Trading. *Nature,* April 14.

Camerer, C., Babcock, L., Loewenstein, G. and Thaler, R. (1997). Labor Supply of New York City Cabdrivers: One Day at a Time. *The Quarterly Journal of Economics*, 112, 2, 407–441.

Carrazedo, T., Curto, J.D. and Oliveira, L. (2016). The Halloween Effect in European Sectors, *Research in International Business and Finance*, 37, 489–500.

Chiappori, P., Levitt, S. and Groseclose, T. (2002). Testing Mixed-Strategy Equilibria When Players Are Heterogeneous: The Case of Penalty Kicks in Soccer, *American Economic Review*, 92, 1138–1151.

Clarke, R.D. (1946). An Application of the Poisson Distribution, *Journal of the Institute of Actuaries*, 72, 481.

Coates, J.M. and Herbert, J. (2008). Endogenous Steroids and Financial Risk Taking on a London Trading Floor, *Proceedings of the National Academy of Sciences of the United States of America,* 15, 16, 6167–6172.

Collins, A., McKenzie, J. and Vaughan Williams, L. (2019). When Is a Talent Contest Not a Talent Contest? Sequential Performance Bias in Expert Evaluation, *Economics Letters*, 177, April, 94–98.

Connor, M.R. ((2020). How A Game of Rock, Paper, Scissors Won A Muli-Million Art Deal for Christie's. Londonist. 11 November.

Danziger, S., Levav, J. and Avnaim-Pesso, L. (2011). Extraneous Factors in Judicial Decisions, *Proceedings of the National Academy of Sciences of the United States of America,* 108, 17, 6889–6892.

Dohmen, T.J. (2008). The Influence of Social Forces: Evidence from the Behavior of Soccer Referees, *Economic Inquiry*, 46, 3, 411–424.

Dowie, J. (1976). On the Efficiency and Equity of Betting Markets, *Economica*, 43, 170, May, 139–150.

Eckhardt, W. (1997). A Shooting-Room View of Doomsday, *The Journal of Philosophy*, 94, 5, 244–259.

Edwards, A. (2013). Ars Conjectandi Three Hundred Years On. *Significance*, 10, June, 39–41.

Ellenberg, J. (2015). *How Not to Be Wrong*. The Hidden Maths of Everyday Life. London: Penguin Books.

Ellerton, P. (2014). Why Facts Alone Don't Change Minds in Our Public Debates. *The Conversation*, May 13.

Fama, E.F. (1970). Efficient Capital Markets: A Review of Theory and Empirical Work, *The Journal of Finance*, 25, 2, May, 383–417.

Feinstein, A. et al. (1985). The Will Rogers Phenomenon—Stage Migration and New Diagnostic Techniques as a Source of Misleading Statistics for Survival in Cancer, *New England Journal of Medicine*, 312, 25, 1604–1608.

Fenton, N., Neil, M. and Berger, D. (2016). Bayes and the Law, *Annual Review of Statistics and Its Applications*, 3, 51–77.

Forgan, D.H. (2019). *Solving Fermi's Paradox*. Cambridge: Cambridge University Press.

Galton, F. (1907). Vox Populi, *Nature*, 75, March 7, 450–451.

Garrett, T., Paton, D. and Vaughan Williams, L. (2020). Taxing Gambling Machines to Enhance Public and Private Revenue, *Kyklos*, 73, 4, 500–523.

Gauriot, R., Page, L. and Wooders, J. (2016). Nash at Wimbledon: Evidence from Half a Million Serves. Working Paper. SSRN, October 12.

Gigerenzer, G. (2003). *Reckoning with Risk. Learning to Live with Uncertainty*. London: Penguin Books.

Good, I.J. (1995). When Batterer Turns Murderer, *Nature*, 375, 541.

Gordon, K.H. (1921). Group Judgements in the Field of Lifted Weights, *Psychological Review*, 28, 6, November, 398–424.

Gott, J.R. (1993). Implications of the Copernican Principle for Our Future Prospects, *Nature*, 363, 6427, 315–319.

Green, P. (2002). Letter from the President of the RSS to the Lord Chancellor regarding the use of statistical evidence in court cases, Royal Statistical Society, January 23.

Griffith, R.M. (1949). Odds Adjustments by American Horse-Race Bettors, *American Journal of Psychology*, 62, 2, 290–294.

Grossman, S.J. and Stiglitz, J. (1980). On the Impossibility of Informationally Efficient Markets, *American Economic Review*, 70, June, 393–408.

Haggard, K.S. and Witte, H.D. (2010). The Halloween Effect: Trick or Treat, *International Review of Financial Analysis,* 19, 5, 379–387.

Haigh, J. and Vaughan Williams, L. (2008). Index Betting for Sports and Stock Indices. In: D. Hausch and W. Ziemba (eds.), *Handbook of Sports and Lottery Markets*, 357–383. Amsterdam: Elsevier, North-Holland.

Hanson, R. and Oprea, R. (2009). A Manipulator Can Aid Prediction Market Accuracy, *Economica*, 76, 302, 304–314.

Hempel, C.G. (1945a). Studies in the Logic of Confirmation I", *Mind*, 54, 13, 1–26.

Hempel, C.G. (1945b). Studies in the Logic of Confirmation II", *Mind*, 54, 214, 97–121.

Henery, R.J. (1985). On the Average Probability of Losing Bets on Horses with Given Starting Price Odds, *Journal of the Royal Statistical Society. Series A (General)*, 148, 4, 342–349.

Hirshleifer, D. and Shumway, T. (2003). Good Day Sunshine: Stock Returns and the Weather, *The Journal of Finance*, 58, 3, June, 1009–1062.

Hooper, M. (2013). Richard Price, Bayes' Theorem and God. *Significance*, 10, February, 36–39.

Horgan, J. (2012). Science Will Never Explain Why There's Something Rather Than Nothing. *Scientific American*, April 23.

Huggett, N. (2018). Zeno's Paradoxes. *Stanford Encyclopedia of Philosophy*, June 11.

Hume, D. (2008). *An Inquiry Concerning Human Understanding. Oxford World's Classics*. Oxford: Oxford University Press.

Jacobsen, B. and Zhang, C.Y. (2012). The Halloween Indicator: Everywhere and All the Time. Working Paper. Massey University, University of New Zealand.

Keane, S.M. (1987). *Efficient Markets and Financial Reporting*. Edinburgh: Institute of Chartered Accountants of Scotland.

Kelly, J.L. (1956). A New Interpretation of Information Rate, *Bell System Technical Journal*, 35, 4, 917–926.

Koshelva, O. and Kreinovich, V. (2018). *How to Explain the Results of the Richard Thaler's 1997 Financial Times Contest*. Department Technical Reports (CS). 1197. University of Texas at El Paso.

Krueger, T.M. and Kennedy, W.F. (1990). An Examination of the Super Bowl Stock Market Predictor, *The Journal of Finance*, 45, 2, 691–697.

Kucharski, A. (2016). *The Perfect Bet. How Science and Maths Are Taking the Luck Out of Gambling*. London: Profile Books Ltd.

Lamont, O.A. and Thaler, R.H. (2003a). Anomalies: The Law of One Price in Financial Markets, *Journal of Economic Perspectives*, 17, 4, 191–202.

Lamont, O.A. and Thaler, R. H. (2003b). Can the Market Add and Subtract? Mispricing in Tech Stock Carve-outs. *Journal of Political Economy*, 111, 2, 227–268.

Lee, M. and King, B. (2017). Bayes' Theorem: The Maths Tool We Probably Use Every Day. But What Is It? *The Conversation*, April 23.

Leslie, J. (1998). *The End of the World: The Science and Ethics of Human Extinction*.London: Routledge.

Levitt, S.D. and Dubner, S.J. (2015). *When to Rob a Bank*. London: Penguin Books.

Lewis, M.A. (2020). Bayes' Theorem and Covid-19 Testing. *Significance*. April 22.

Malkiel, B. (2003). The Efficient Market Hypothesis and Its Critics. CEPE Working Paper 91. Princeton University.

Mantonakis, A., Rodero, P., Lesschaeve, I. and Hastie, R. (2009). Order in Choice: Effects of Serial Position on Preferences, *Psychological Science*, 20, 1, 1309–1312.

Matthew 20: 1–20. Gospel of Matthew. New Testament, The Bible. Parable of the Workers in the Vineyard.

Mlodinow, L. (2009). *The Drunkard's Walk. How Randomness Rules Our Lives*. London: Penguin Books.

Morgenstern, O. and von Neumann, J. (1947). *The Theory of Games and Economic Behavior*. Princeton: Princeton University Press.

Moskowitz, C. (2016). Are We Living in a Computer Simulation? *Scientific American*, April 7.

Moskowitz, T.J. and Wertheim, L.J. (2011). *Scorecasting*. New York: Random House.

Nash, J. (1950). Equilibrium Points in *n*-Person Games, *Proceedings of the National Academy of Sciences of the United States of America*, 36, 1, 48–49.

Nash, J. (1951). Non-Cooperative Games, *The Annals of Mathematics*, 54, 2, 286–295.

Nevill, A.M., Balmer, N.J. and Williams, A.M. (2002). The Influence of Crowd Noise and Experience Upon Refereeing Decisions in Football, *Psychology of Sport and Exercise*, 3, 4, 261–272.

Newcomb, S. (1881). Note on the Frequency of Use of the Different Digits in Natural Numbers, *American Journal of Mathematics*, 4, 1, 39–40.

New Scientist. (2015). *Chance. The Science and Secret of Luck, Randomness and Probability*. In: M. Brooks (ed.). London: Profile Books Ltd.

Nozick, R. (1969). Newcomb's Problem and Two Principles of Choice. In: N. Rescher et al. (eds.), *Essays in Honor of Carl G. Hempel*. New York: Springer, 114–146.

Office for National Statistics. (2021). Deaths Involving COVID-19 by Vaccination Status, England: Deaths Occurring between 2 January and 24 September 2021.

Orkin, M. (1991). *Can You Win? The Real Odds for Casino Gambling, Sports Betting, and Lotteries. With a Chapter on Prisoner's Dilemma*. New York: W.H. Freeman and Company.

Pacioli, L. (1494). Summa de arithmetica, geometria, proportioni et proportionalita [Summary of Arithmetic, Geometry, Proportions and Proportionality]. Venice: Paganini.

Parfit, D. (1998a). Why Anything? Why This? Part 1, *London Review of Books*, 20, 2, January 22, 24–27.

Parfit, D. (1998b). Why Anything? Why This? Part 2, *London Review of Books*, 20, 3, February 5, 22–25.

Pascal, B. (1665). *Traité du triangle arithmétique*. Cambridge: Cambridge University Library.

Pascal, B. (1670). *Pensées* [Thoughts], 2nd. Edn. Paris: Chex Guillaume Desprez.

Paton, D., Siegel, D. and Vaughan Williams, L. (2002a). A Policy Response to the E-Commerce Revolution: The Case of Betting Taxation in the UK, *The Economic Journal*, 12, 480, F296–F314.

Paton, D., Siegel, D. and Vaughan Williams, L. (2002b). Gambling Taxation: A Comment, *The Australian Economic Review*, 34, 4, 437–440.

Paton, D., Siegel, D. and Vaughan Williams, L. (2004). Taxation and the Demand for Gambling: New Evidence from the United Kingdom, *National Tax Journal*, 57, 4, 847–861.

Paton, D., Siegel, D. and Vaughan Williams, L. (2009). The Growth of Gambling and Prediction Markets: Economic and Financial Implications, *Economica*, 76, 302, 219–224.

Paton, D., Siegel, D. and Vaughan Williams, L. (2010). Gambling, Prediction Markets and Public Policy, *Southern Economic Journal*, 76, 4, 878–883.

Paton, D. and Vaughan Williams, L. (1998). Do Betting Costs Explain Betting Biases? *Applied Economics Letters*, 5, 5, 333–335.

Paton, D. and Vaughan Williams, L. (2005). Forecasting Outcomes in Spread Betting Markets: Can Bettors Use 'Quarbs' to Beat the Book?, *Journal of Forecasting*, 24, 2, 139–154.

Poisson, S.D. (1837). *Probabilité des jugements en matière criminelle et en matière civile, précédées des règles générales du calcul des probabilités* [Research on the Probability of Judgments in Criminal and Civil Matters]. Paris: Bachelier.

Pope, D.G. and Schweitzer, M.E. (2011). Is Tiger Woods Loss Averse? Persistent Bias in the Face of Experience, Competition and High Stakes, *American Economic Review*, 101, 1, 129–157.

Poundstone, W. (2005). *Fortune's Formula. The Untold Story of the Scientific System That Beat the Casinos and Wall Street.* New York: Hill and Wang.

Poundstone, W. (2015). *How to Predict the Unpredictable: The Art of Outsmarting Almost Everyone.* London: Oneworld Publications.

Price, J. and Wolfers, J. (2014). Right-Oriented Bias: A Comment on Roskes, Sligte, Shalvi, and De Dreu (2011), *Psychological Science*, 25, 11, 2109–2111.

Puga, J., Krzywinski, N. and Altman, N. (2015). Points of Significance: Bayes' Theorem, *Nature Methods*, 12, 4, April, 277–278.

Reade, J. and Vaughan Williams, L. (2019). Polls to Probabilities: Comparing Prediction Markets and Opinion Polls, *International Journal of Forecasting*, 35, 1, 336–350.

Reade, J., Singleton, C. and Vaughan Williams, L. (2020). Betting Markets for English Premier League Results and Scorelines: Evaluating a Forecasting Model, *Economic Issues*, 25, 1, 87–106.

Robertson, B., Vignaux, G.A. and Berger, C.E.H. (2016). *Interpreting Evidence: Evaluating Forensic Science in the Courtroom*, 2nd. Edn. Chichester: Wiley.

Roskes, M., Sligte, D., Shalvi, S. and De Dreu, C.K.W. (2011). The Right Side? Under Time Pressure, Approach Motivation Leads to Right-Oriented Bias, *Psychological Science*, 22, 11, 1403–1407.

Salop, S.C. (1987). Evaluating Uncertain Evidence with Sir Thomas Bayes: A Note for Teachers, *Economic Perspectives*, 1, 1, Summer, 155–160.

Saunders, E. (1993). Stock Prices and Wall Street Weather, *American Economic Review*, 83, 5, 1337–1345.

Saville, B., Stekler, H. and Vaughan Williams, L. (2011). Do Polls or Markets Forecast Better? Evidence from the 2010 US Senate Elections, *The Journal of Prediction Markets*, 5, 3, 64–74.

Schmidt, B. and Clayton, R. (2017). Super Bowl Indicator and Equity Markets: Correlation Not Causation, *Journal of Business Inquiry*, 17, 2, 97–103.

Schrödinger, E. (1935). *Die gegenwärtige Situation in der Quantenmechanik* [The Present Situation in Quantum Mechanics], *Naturwissenschaften*, 23, 48, 807–812.

Science Buddies. (2012). Probability and the Birthday Paradox, *Scientific American*, March 29.

Shakespeare, W. (2018a). *The Merchant of Venice.* M.M. Mahood (ed.). (The New Cambridge Shakespeare). Cambridge: Cambridge University Press.

Shakespeare, W. (2018b). *Othello.* N. Sanders (ed.). (The New Cambridge Shakespeare). Cambridge: Cambridge University Press.

Shostak, S. (2019). 'Zoo Hypothesis' May Explain Why We Haven't Seen Any Space Aliens. *NBC News*, March 31.

Significance. 2021. A School Named Bayes. June, 18, 3.

Silver, N. (2012). *The Signal and the Noise: The Art and Science of Prediction.* London: Allen Lane.

Skorupski, W.P. and Wainer, H. (2015). The Bayesian Flip. Correcting the Prosecutor's Fallacy. *Significance*, 12, August, 16–20.

Smart, J.M. (2012). The Transcension Hypothesis: Sufficiently Advanced Civilizations Invariably Leave Our Universe, and Implications for METI and SETI, *Acta Astronautica*, 78, September–October, 55–68.

Smith, M.A., Paton, D. and Vaughan Williams, L. (2005). An Assessment of Quasi-Arbitrage Opportunities in Two Fixed-Odds Horse-Race Betting Markets. In: L. Vaughan Williams (ed.), *Information Efficiency in Financial and Betting Markets.* Chapter 4. Cambridge: Cambridge University Press, 159-171.

Smith, M.A., Paton, D. and Vaughan Williams, L. (2006). Market Efficiency in Person-to-Person Betting, *Economica*, 73, 292, 673–689.

Smith, M.A., Paton, D. and Vaughan Williams, L. (2009). Do Bookmakers Possess Superior Skills to Bettors in Predicting Outcomes? *Journal of Economic Behavior and Organization*, 71, 2, 539–549.

Smith, M.A. and Vaughan Williams, L. (2010). Forecasting Horse Race Outcomes: New Evidence on Odds Bas in UK Betting Markets, *International Journal of Forecasting*, 26, 3, 543–550.

Stangroom, J. (2009). *Einstein's Riddle. Riddles, Paradoxes and Conundrums to Stretch Your Mind*. London: Bloomsbury Publishing Plc.

Stewart, I. (2019). *Do Dice Play God? The Mathematics of Uncertainty*. London: Profile Books Ltd.

Stiglitz, J.E. (1981). Information and the Change in the Paradigm of Economics. *Nobel Prize Lecture*, December 8.

Stovall, R. (1989). The Super Bowl Predictor, *Financial World*, 158, 72.

Summers, L.H. (1985). On Economics and Finance, *The Journal of Finance*, 40, 3, 633–635.

Surowiecki, J. (2004). *The Wisdom of Crowds. Why the Many Are Smarter Than the Few and How Collective Wisdom Shapes Business, Economies, Societies, and Nations*. New York: Doubleday.

Talwalkar, P. (2014). *The Joy of Game Theory. An Introduction to Strategic Thinking*. CreateSpace Independent Publishing Platform.

Talwalkar, P. (2015). *Math Puzzles Volume 1. Classic Riddles and Brain Teasers in Counting, Geometry, Probability, and Game Theory*. CreateSpace Independent Publishing Platform.

Talwalkar, P. (2015). *Math Puzzles Volume 1. Even More Riddles and Brain Teasers in Counting, Geometry, Probability, and Game Theory*. CreateSpace Independent Publishing Platform.

Talwalkar, P. (2015). *Math Puzzles Volume 2. More Riddles and Brain Teasers in Counting, Geometry, Probability, and Game Theory*. CreateSpace Independent Publishing Platform.

Talwalkar, P. (2015). *40 Paradoxes in Logic, Probability, and Game Theory*. CreateSpace Independent Publishing Platform.

Talwalkar, P. (2016). *The Irrationality Illusion. How to Make Smart Decisions and Overcome Bias*. CreateSpace Independent Publishing Platform.

Tetlock, P. and Gardner, D. (2016). *Superforecasting: The Art and Science of Prediction*. London: Random House.

Thaler, R. (1997a). Competition. *Financial Times*, May 9, Sec. 1, 29.

Thaler, R. (1997b). Giving Markets a Human Dimension. *Financial Times*, June 16, Sec. 6, 2–5.

The Council of the Inns of Court and the Royal Statistical Society (2019). Statistics and probability for advocates: Understanding the use of statistical evidence in court and tribunals.

Thorp, E.O. (1966). *Beat the Dealer: A Winning Strategy for the Game of Twenty-One*. New York: Random House USA.

Tijms, H. (2019). Surprises in Probability—Seven Short Stories. Boca Raton: CRC Press. Taylor & Francis Group.

Treynor, J. (1987). Market Efficiency and the Bean Jar Experiment, *Financial Analysts Journal*, 43, 50–53.

Trotta, R. (2014). Is the All-There-Is All There Is? *Significance*, June, 69–71.

Tversky, A. and Kahneman, D. (1982). Evidential Impact of Base Rates. In: D. Kahneman, P. Slovic and A. Tversky (eds.), Judgment under Uncertainty: Heuristics and Biases. Cambridge: Cambridge University Press, 153–160.

Vaughan Williams, L. (1999). Information Efficiency in Betting Markets: A Survey, *Bulletin of Economic Research*, 51, 1, 1–39.

Vaughan Williams, L. (2000). Can Forecasters Forecast Successfully? Evidence from UK Betting Markets, *Journal of Forecasting*, 19, 6, 505–513.

Vaughan Williams, L. (2003a). *The Economics of Gambling*. London: Routledge.

Vaughan Williams, L. (2003b). *Betting to Win: A Professional Guide to Profitable Betting*. Oldcastle Books Ltd.

Vaughan Williams, L. (2005). *Information Efficiency in Financial and Betting Markets*. L. Vaughan Williams (ed.). Cambridge: Cambridge University Press.

Vaughan Williams, L. (2011). *Prediction Markets: Theory and Applications*. L. Vaughan Williams (ed.). Abingdon: Routledge International Studies in Money and Banking.

Vaughan Williams, L. (2012). *The Economics of Gambling and National Lotteries. The International Library of Critical Writings in Economics*. Cheltenham: Edward Elgar.

Vaughan Williams, L. (2014). The Churchill Betting Tax, 1926-30: A Historical and Economic Perspective, *Economic Issues*, 19, 2, 21–38.

Vaughan Williams, L. (2015). Forecasting the Decisions of the US Supreme Court: Lessons from the 'Affordable Care Act' Judgment, *The Journal of Prediction Markets*, 9, 2, 64–78.

Vaughan Williams, L. (2018). Written Evidence (PPD0024). House of Lords Political Polling and Digital Media Committee. January 16. https://data.parliament.uk/writtenevidence/committeeevidence.svc/evidencedocument/political-polling-and-digital-media-committee/political-polling-and-digital-media/written/72373.pdf

Vaughan Williams, L. and Paton, D. (1997a). Why Is There a Favourite-Longshot Bias in British Racetrack Betting Markets? *The Economic Journal*, 107, 150–158.

Vaughan Williams, L. and Paton, D. (1997b). Does Information Efficiency Require a Perception of Information Inefficiency? *Applied Economics Letters*, 4, 10, 615–617.

Vaughan Williams, L. and Paton, D. (1998). Why Are Some Favourite-Longshot Biases Positive and Others Negative? *Applied Economics*, 30, 11, 1505–1510.

Vaughan Williams, L. and Paton, D. (2015). Forecasting the Outcome of Closed-Door Decisions: Evidence from 500 Years of Betting on Papal Conclaves, *Journal of Forecasting*, 34, 5, 391–404.

Vaughan Williams, L. and Reade, J. (2015). Forecasting Elections, *Journal of Forecasting*, 35, 4, 308–328.

Vaughan Williams, L. and Reade, J. (2016). Prediction Markets, Social Media and Information Efficiency, *Kyklos*, 69, 3, 518–556.

Vaughan Williams, L. and Siegel, D. (2013). *The Oxford Handbook of the Economics of Gambling*. New York: Oxford University Press.

Vaughan Williams, L. and Stekler, H. (2010). Sports Forecasting, *International Journal of Forecasting*, 26, 3, 445–447.

Vaughan Williams, L. and Vaughan Williams, J. (2009). The Cleverness of Crowds, *The Journal of Prediction Markets*, 3, 3, 45–47.

Vaughan Williams, L., Sung, M. and Johnson, J.E.V. (2019). Prediction Markets: Theory, Evidence and Applications, *International Journal of Forecasting*, 35, 1, 266–270.

Vaughan Williams, L., Sung, M., Fraser-Mackenzie, P., Peirson, J. and Johnson, J.E.V. (2016). Towards an Understanding of the Origins of the Favourite-Longshot Bias: Evidence from Online Poker, a Real-World Natural Laboratory, *Economica*, 85, 338, 360–382.

Vogel, C. (2005). Rock, Paper, Payoff: Child's Play Wins Auction House an Art Sale. New York Times, April 29.Von Bortkiewicz, L. (1898). *Das Gesetz der kleinen Zahlen* [The Law of Small Numbers]. Leipzig: B.G. Teubner.

Vos Savant, M. (1990). Ask Marilyn Column. *Parade*, September 9.

Walker, M. and Wooders, J. (2001). Minimax Play at Wimbledon, *American Economic Review*, 91, 1521–1538.

Waniek, M., Niescieruk, A., Michalak, T. and Rahwan, T. (2015). Spiteful Bidding in the Dollar Auction. *Proceedings of the Twenty-Fourth Internationl Joint Conference on Artificial Intelligence*.

Wason, P.C. (1966). Reasoning. In: B.M. Foss (ed.). *New Horizons in Psychology.* Vol. 1. Harmondsworth: Penguin, 135–151.

Wason, P.C. (1968). Reasoning About a Rule, *Quarterly Journal of Experimental Psychology*, 20, 3, 273–281.

Zhang, C.Y. and Jacobsen, B. (2021). The Halloween Indicator: Sell in May and Go Away: Everywhere and All the Time, *Journal of International Money and Finance*, 110, February, 1–49.

Index

Printed in the United States
by Baker & Taylor Publisher Services